WEYERHAEUSER ENVIRONMENTAL BOOKS

Paul S. Sutter, Editor

WEYERHAEUSER ENVIRONMENTAL BOOKS explore human relationships with natural environments in all their variety and complexity. They seek to cast new light on the ways that natural systems affect human communities, the ways that people affect the environments of which they are a part, and the ways that different cultural conceptions of nature profoundly shape our sense of the world around us. A complete list of the books in the series appears at the end of this book.

PEOPLE OF THE ECOTONE

Environment and Indigenous Power at the Center of Early America

ROBERT MICHAEL MORRISSEY

UNIVERSITY OF WASHINGTON PRESS
Seattle

People of the Ecotone is published with the assistance of a grant from the Weyerhaeuser Environmental Books Endowment, established by the Weyerhaeuser Company Foundation, members of the Weyerhaeuser family, and Janet and Jack Creighton.

A subvention from the Mellon Foundation and the Humanities Research Institute at the University of Illinois supported the publication of this book.

Composed in Minion Pro, typeface designed by Robert Slimbach.

UNIVERSITY OF WASHINGTON PRESS
uwapress.uw.edu

LIBRARY OF CONGRESS CATALOGING-IN-PUBLICATION DATA
Names: Morrissey, Robert Michael, author.
Title: People of the ecotone : environment and indigenous power at the center of Early America / Robert Michael Morrissey.
Other titles: Environment and indigenous power at the center of Early America
Description: Seattle : University of Washington Press, 2022. | Series: Weyerhaeuser environmental books | Includes bibliographical references and index.
Identifiers: LCCN 2022015078 (print) | LCCN 2022015079 (ebook) | ISBN 9780295750873 (hardcover) | ISBN 9780295750880 (paperback) | ISBN 9780295750897 (ebook)
Subjects: LCSH: Indians of North America—Middle West—History. | Illinois Indians—History. | Fox Indians—History. | Prairie ecology—Middle West—History. | Ecotones—Middle West—History. | Human ecology—Middle West—History. | French—Middle West—History. | Middle West—History.
Classification: LCC E78.M67 M65 2022 (print) | LCC E78.M67 (ebook) | DDC 977/.01—dc23/eng/20220615
LC record available at https://lccn.loc.gov/2022015078
LC ebook record available at https://lccn.loc.gov/2022015079

♾ This paper meets the requirements of ANSI/NISO Z39.48-1992 (Permanence of Paper).

Here you ordinarily begin to see the buffalo.

—PIERRE CHARLES DE LIETTE,
writing from the Illinois Country, ca. 1690

CONTENTS

FOREWORD

Ecotone History

PAUL S. SUTTER

TWO DECADES AGO, IN THE PROLOGUE TO HIS INFLUENTIAL REORIENTATION of early American history, *Facing East from Indian Country*, Daniel K. Richter recounted an epiphany he had while gazing east through the Gateway Arch from the window of a St. Louis hotel room. The arch celebrates the territorial westering of the United States, and St. Louis's place in that story, perched as it is on the Mississippi River just below its confluence with the Missouri. A monument meant to be viewed from the east, the arch embodied, for Richter, the traditional geographical perspective of early American history, whose westward gaze had long defined the eastern colonies as the center of the action. The problem with that narrative was that Indigenous peoples stood on the edge of history looking in, marginal and without agency. Richter asked: What if we reverse the polarity and narrate early American history by facing east, as he was doing that evening? What would we see if we stood atop an imaginary Gateway Arch—or, better yet, atop Cahokia's towering temple mound—from the sixteenth through the eighteenth centuries? The answer is a vast geography of Indigenous power and historical dynamism. Centering early American history in the heart of the continent would highlight the persistent capacity of diverse Indigenous peoples to make their own histories according to their own logics. Historians of early America, Richter realized, were ignoring most of early America and, in the process, misconstruing the entire continent's history.

Sitting just upstream from St. Louis is another confluence, the mixing place of the Mississippi and Illinois Rivers. If one had traveled up the Illinois to its headwaters in the late seventeenth century, one would have encountered the Grand Village of the Kaskaskias and its surrounding settlements, a conurbation of the Illinois and allied peoples, perhaps twenty thousand strong, that was, at that time, one of the largest population centers in North America. It was likely larger than Boston, and New York, and Philadelphia. It was from just such a place that Richter encouraged us to reimagine the history of early America, and that is just what Robert Morrissey has done in this powerfully revisionist and richly layered study, *People of the Ecotone*. Morrissey's book enthusiastically partakes in the reorientation that Richter envisioned; he gives us a history of the Illinois, their allies, and their rivals that is driven by Indigenous motivations, a history in which fluorescence is the dominant narrative arc and European forces—disease, market penetration, imperial rivalry, and settler colonialism—are both peripheralized and reconceptualized. That alone makes this an important book. But Morrissey's story of Indigenous power in the heartland does something else as well, something that requires its own perspectival and methodological reorientation.

If one had been able to achieve a commanding view from the ruins of Cahokia in the centuries before the United States became a nation, one would have noticed something else: a vast peninsula of tallgrass prairie that pushed eastward from the Great Plains and covered most of what is today the state of Illinois plus several states to the west. Once one of the continent's dominant biomes, the tallgrass prairie is now almost entirely gone, swallowed up by an industrial monotony of soy and corn. Its erasure has resulted in what Morrissey calls a "collective amnesia." We have lost a sense of the ecosystem itself. The flat and seemingly featureless contemporary Midwest has, incorrectly, encouraged us to imagine the precontact landscape as similarly flat and featureless, a primordial sea of grass that stretched beyond the horizon. The reality was far more complex, for the tallgrass region was a diverse mosaic of prairies, wetlands, and woodlands, a pulsating ecotone between the thick forests of the east and the expansive grasslands of the west. The tallgrass prairie was not, according to Morrissey, a timeless natural setting, a static stage upon which history played out. It had its own history of change over time, its own cutting edge. *People of the Ecotone* thus does for the bioregion what historians of Indigenous America have done for the peoples of Indian Country: portrays it as historically dynamic and powerful in its own right. More than that, though, Morrissey urges us to remember that Indigenous power was not placeless. It existed in dynamic relation with environments that had their own histories. For the Illinois of the Grand Village of the Kaskaskias, and for the others who came to the region to take advantage of its opportunities, their collective history *took place* within the tallgrass prairie ecotone.

By weaving together the history of Indigenous peoples and the history of the ecotone—indeed, by seeing them as one and the same—*People of the Ecotone* fundamentally changes how we understand what was happening on the prairie edge during the sixteenth, seventeenth, and early eighteenth centuries. Most historians have characterized the Grand Village of the Kaskaskias not as a manifestation of Indigenous power but as a symptom of crisis; for them it looked like a vast refugee camp for people fleeing the violence of the so-called Beaver Wars of the Great Lakes region and a defensive sign of Indigenous devastation and decline, the beginning of the end for the Illinois and their way of life. For Morrissey it was the culmination of a trajectory that began decades, even centuries, earlier. That trajectory in turn led to the Fox Wars of the early decades of the 1700s, a conflict whose violence historians have, until now, struggled to see through Indigenous eyes. As many historians have done, Morrissey rejects French depictions of this violence as somehow an essential feature of Indigenous life in the region. He also disabuses us of the tendency to see it merely as a reaction to colonial upheaval or as the destruction of precontact cultures and stabilities. To understand the violence, he insists, we need to understand ecotone history.

In Morrissey's skillful hands, ecotone history is something new, an amalgamation of environmental history and borderlands history that is more than the sum of its parts. The tallgrass prairie was certainly a borderland in the traditional sense of that term: a place where different peoples met, mixed, and came into conflict. But it was also an ecological borderland whose edge effects bled across the boundaries between natural and human history. And while this book certainly fits well within the field of environmental history, it also chafes against an ontological separation that many environmental historians too often fall back on in justifying their field: that *nature* matters to *history*. In ecotone history there is no nature outside of history; indeed, it is impossible to pry those two categories apart.

People of the Ecotone begins with a deep history of the tallgrass prairie, a Holocene biome constantly in the making. This is not a scene setting but an introduction to a suite of cast members in the drama that follows. In the

1500s, as a result primarily of climatic oscillations, large numbers of bison entered the ecotone, sensing opportunity there, and humans did the same, reshaping the ecotone with fire to manage the continuing flux. Humans and bison together shaped their fortunes in a process that environmental humanities scholars have come to call multispecies worldmaking. From this contingent moment in ecotone history, Morrissey argues, there emerged a new energy regime: a unique, and thus far neglected, form of pedestrian bison hunting that expressed itself in Indigenous physiology, economy, gender relations, and settlement patterns well before the heavy hand of European contact. This assemblage of forces, in turn, shaped particular kinds of violence.

Key to Morrissey's analysis is a bold reinterpretation of how the people of the ecotone adapted traditional mourning wars—military forays driven by the desire to replace deceased kin lost in battle—into novel patterns of conflict, slavery, and diplomacy specific to the bioregion. By the late 1600s, remade by their commensal relationship with bison, the people of the ecotone committed themselves to a more intensive form of pedestrian bison hunting and hide processing that eventually brought them into the Grand Village and its satellite communities amid large concentrations of bison that were themselves products of history. When we see it as a manifestation of ecotone history, the Grand Village does not appear to be a defensive reaction to exogenous forces, the product of cultures shattered and reconstituted by the coming of colonialism. Viewed from what Morrissey evocatively calls "the edge at the center," the Grand Village appears as an Indigenous achievement made possible by historically specific forces and opportunities.

The Grand Village was short-lived. The entangled dynamism of ecotone history shifted yet again, and so did the arrangements of Indigenous peoples there. The violence of the Fox Wars would be one result. Again, it has been easy to see this shift as being propelled by outside forces, and those forces certainly played a role. But Morrissey holds fast to an ecotonal perspective that renders the story still fully in the hands of the region's Indigenous peoples. "Indigenous trajectories and multispecies entanglements," he concludes, "determined *how colonialism came to matter.*"

I cannot do justice in this brief foreword to Morrissey's rich, layered, and utterly compelling narrative, and I would not want to spoil its most revealing moments. Suffice it to say that even those who come to this book knowing little about this story or how others have told it will come away understanding history anew. *People of the Ecotone* brings a glass-bottom boat to a rippling sea of soy and corn, a region remade by industrial monoculture, to reveal a remarkable history that is still there, beneath the surface. To invoke his wonderfully evocative double entendre, Robert Morrissey has found a whole new way of understanding early American history "hiding in the tallgrass."

ACKNOWLEDGMENTS

I THANK THE CENTER FOR ADVANCED STUDY AT THE UNIVERSITY OF Illinois and the National Endowment for the Humanities (grant number FA-58336-15) for the original fellowships that supported this project. Thanks also to the Illinois Program for Research in the Humanities (now the Humanities Research Institute) and the Mellon Foundation for naming me the IPRH-Mellon Faculty Fellow in Environmental Humanities for 2018–20, during which I wrote much of this book.

Many people helped me prepare to write this book. I'm especially grateful to Tim Newfield, Lee Mordechai, and John Haldon for organizing the Climate Change and History Workshops at Princeton University. A workshop on dendrochronology, led by Ulf Büntgen and Jürg Luterbacher, and one on palynology, led by Neil Roberts and Warren Eastwood, expanded the scope of this project. Closer to home, I received essential help from Eric Freyfogle, Bruce Hannon, the late Elisabeth M. Hanson, Fran Harty, Kim Curtis, Jason Thomason, Jamie Ellis and John (Jack) White. Chris Widga hosted me for an afternoon at the Illinois State Museum, and Terry Martin shared much expertise as well. Tom Emerson and Duane Esarey, both formerly of the Illinois State Archaeological Survey, helped me dig into archaeology. I could not be more grateful to Lenville Stelle for all his generosity and expertise. Thanks to him and Vicki Smith for enabling my exploration at ML-6 (the Fox Fort). On a very different plane, I'm thankful to the painter Philip Juras for conversations about prairie history, and for generous permission to use his work on the cover illustration of this book.

Many fellow historians helped to shape this book. Dave Edmunds's visit to Illinois in 2011 was my first occasion to visit the Smith farm. The organizers and participants at the Early American Environments Workshop at the Huntington Library (sponsored by the *William and Mary Quarterly* and Early Modern Studies Institute) helped me at a formative moment. At Yale University my former teachers Jay Gitlin, John Mack Faragher, and George Miles, along with Alejandra Dubcovsky, Ned Blackhawk, and Paul Sabin, gave helpful feedback on my early ideas. The Omohundro Institute at the College of William and Mary let me highlight this little corner of "vast early America" in an annual lecture series of that name. Over the ensuing years many other historian friends have heard me discuss ideas and shared their insights, especially John Brooke, Gerry Cadava, Katie Cangany, Jon Coleman, John Demos, Robert Englebert, Blake Gilpin, Kate Grandjean, Patrick Griffin, Richard Gross, Craig Howard, Ben Johnson, Charles Keith, Jacob Lee, Jake Lundberg, Karen Marrero, John Nelson, Chris Parsons, Josh Piker, Carolyn Prodruchny, Jim Rice, Brett Rushforth, Sam Schaffer, Susan Sleeper-Smith, Andrew Sturtevant, and Sophie White. I want to especially thank Liz Ellis, George Ironstrack, David Costa, Cam Shriver, Scott Shoemaker, and other members of an ongoing collaborative project from which I have learned so much about subjects in this book.

This book is fundamentally shaped by the IPRH-Mellon Initiative in Environmental Humanities at the University of Illinois. I thank members of our working group, including John Levi Barnard, Clara Bosak-Schroeder, Carolyn Fornoff, Heidi Hurd, Jamie Jones, Brett Kaufman, Bob Markley, Lindsay Marshall, Mary Pat McGuire, Rebecca Oh, David Sepkoski, Rod Wilson, and Gillen Wood. The best part of that project was working with students, especially Sarah Gediman, Alaina Bottens, Amanda Watson, April Wendling, Juan Luna Nuñez, Clara Pokorny, Doug Jones, Jessica Landau, Alexis Paterson, and Samantha Good. Leah Aronowsky and Pollyanna Rhee were amazing collaborators and shared many insights. Throughout our program no one was more helpful to our group than Nancy Castro, along with the super-generous Jenna Zieman, Stephanie Uebelhoer, and Jennifer Sturm. Antoinette Burton has been an incredible mentor and generous friend to me over many years.

A workshop by the National Center for Faculty Development and Diversity helped me at a crucial time, and I thank especially Rachel Navarro, Ciaran Trace, Kafi Kumasi, and Eboni Baugh for their insights. Provost Amy Santos at UIUC made funding available for this experience.

I am lucky to be at the University of Illinois. I am grateful to Amanda Ciafone, Clare Crowston, Jenny Davis, Jane Desmond, Sam Froiland, Dan Gilbert, Matt Gilbert, Marc Hertzman, Ikuko Asaka, Rana Hogarth, Kristin Hoganson, David Lehman, Riley Liska, John Lynn, Mauro Nobili, Dana Rabin, Eric Toups, and the Premodern Reading Group, and many others, who have helped me along. Craig Koslowsky, Carol Symes, and Maria Gillombardo helped me secure fellowships for this project, and Craig has been a wise mentor through the writing process. I also want to thank the many librarians at the University of Illinois who worked to make materials available throughout the disruptions of the pandemic. Fred Hoxie is a true friend and extraordinary mentor, and I cannot thank him adequately for all the help he has given me.

Paul Sutter and Andrew Berzanskis have been model editors, and I am grateful for their expertise and generosity. Thanks also to the anonymous reviewers for the University of Washington Press, and Joeth Zucco and Ann Baker for their work turning the manuscript into a book. Sally Brown made a very nice index.

This book has a lot to do with running, history, and the prairie landscape. Members of the Carleton College Men's Cross Country running team 1995–99 nurtured my early enthusiasms about all three. Closer to the present, members of my extended family have encouraged this project for many years, and I am grateful. Most importantly, I thank Charlie, Julia, Maeve, and above all Haley, for love and support.

Portions of this book were previously published as "Bison Algonquians: Cycles of Violence and Exploitation in the Mississippi Valley Borderlands," *Early American Studies: An Interdisciplinary Journal—Special Issue: Environment* 13, no. 2 (2015): 309–40; "Climate, Ecology and History in North America's Tallgrass Prairie Borderlands," *Past & Present* 245, no. 1 (November 1, 2019): 39–77; "The Power of the Ecotone: Bison, Slavery, and the Rise and Fall of the Grand Village of the Kaskaskia," *Journal of American History* 102, no. 3 (December 1, 2015): 667–92. Used with permission.

A NOTE ON TERMINOLOGY AND NAMES

The word "Illinois" refers properly to an Indigenous language rather than a people. Understanding that the language was shared widely among the loosely organized Indigenous villages of what they called "Illinois Country," French travelers in the region employed this word as an ethnonym instead of the one that those groups used for themselves: Inoca. Since this latter term thus almost never appears in historical sources and is generally not used among Indigenous people in the present, in this book I use Illinois to name the people because of its continued salience. When appropriate I also name the various ethnic groups or "familles" of Illinois speakers specifically, groups that included the Kaskaskias (Kaahkaahkia), the Peorias (Peewaalia), the Moingwenas, the Cahokias, the Chepoussas, the Chinkoas, the Coiracoentanons, the Espiminkias, the Tapouaros, the Mitchigameas, and the Tamaroas. The Miamis (or Myaamias), along with the Weas (Waayaahtanwas) and the Piankeshaws (Peeyankihsias), also speak the Miami-Illinois language, although all were separate groups during the period this book recounts.

French sources almost always refer to the Meskwakis as "Renards" (Foxes) or "Outagamis." In this book I use the names Meskwaki(s) and Fox(es) interchangeably.

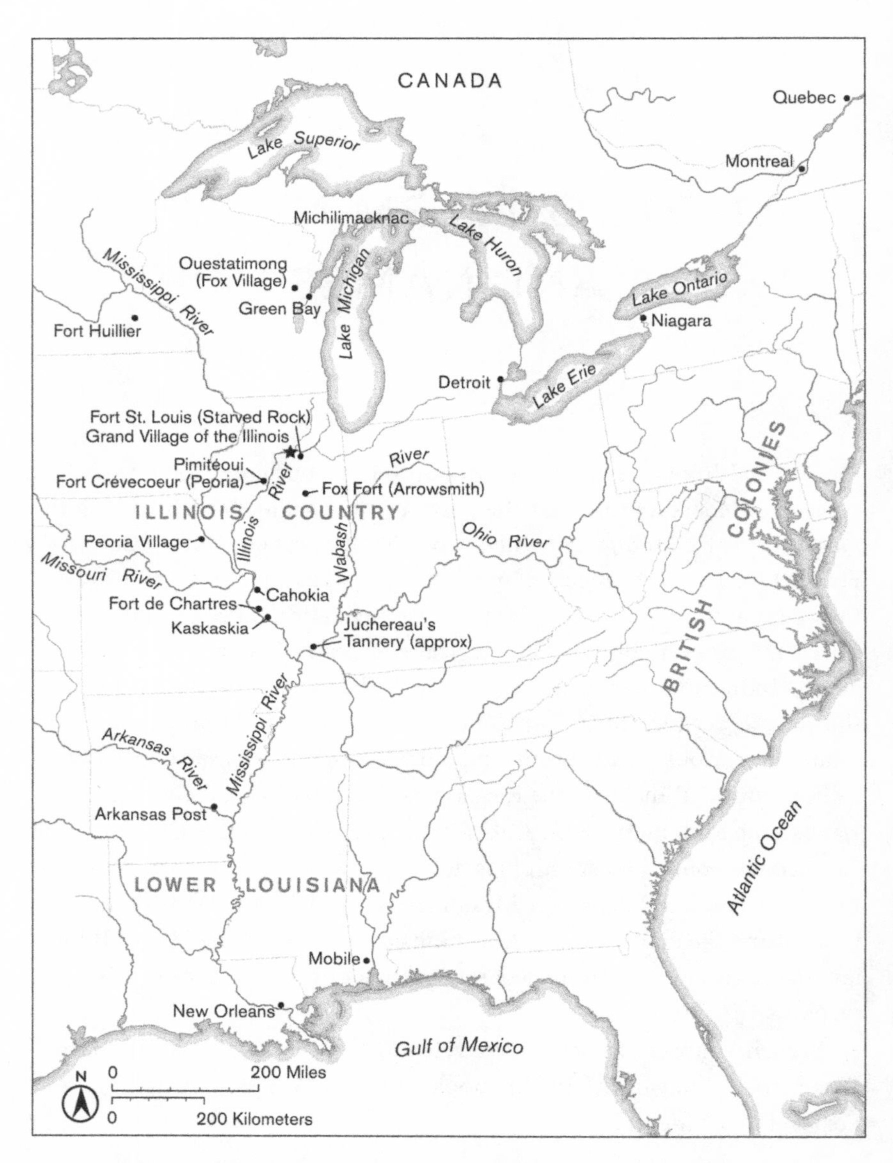

Villages and important locations in the eastern prairie region, 1600–1800. Map by Ben Pease.

PEOPLE OF THE ECOTONE

INTRODUCTION

CONTINUAL WARS AND THE PLACE WHERE THEY LIVED

AN ECOLOGICAL MAP OF THE MIDWEST OF THE UNITED STATES TODAY is relatively boring. If it shows vegetation, the map's variety is minimal, for so much of the present midcontinent is covered by two plants: soy and corn. If the map shows elevations or other geological features, it also presents a similarly uniform aspect of flatness, reflecting a defining characteristic of a landscape that most observers identify with monotony and plainness. This book began with a rather simple, if not simplistic, intuition: an ecological map of the Midwest from several hundred years ago would look very different.

When cartographers attempt to depict the midwestern landscape in its earlier ecological configurations, meaning before industrial agriculture, they reveal a stark and conspicuous fact: the heart of the Midwest once contained not a uniformity but a tremendous and dramatic dichotomy. The center of the midcontinent was a biome-scale transition, the division between woodlands and grasslands, a massive tension zone that included forests and prairies as well as wetlands, savannas, and other assemblages. Over thousands of years, from around 5000 BCE through to 1850 CE, this dynamic transition stretched for hundreds of miles east to west, a triangle with its base running from the eastern edge of the Dakotas south to the Ozarks of northwestern Arkansas and its two sides extending diagonally eastward to a point near

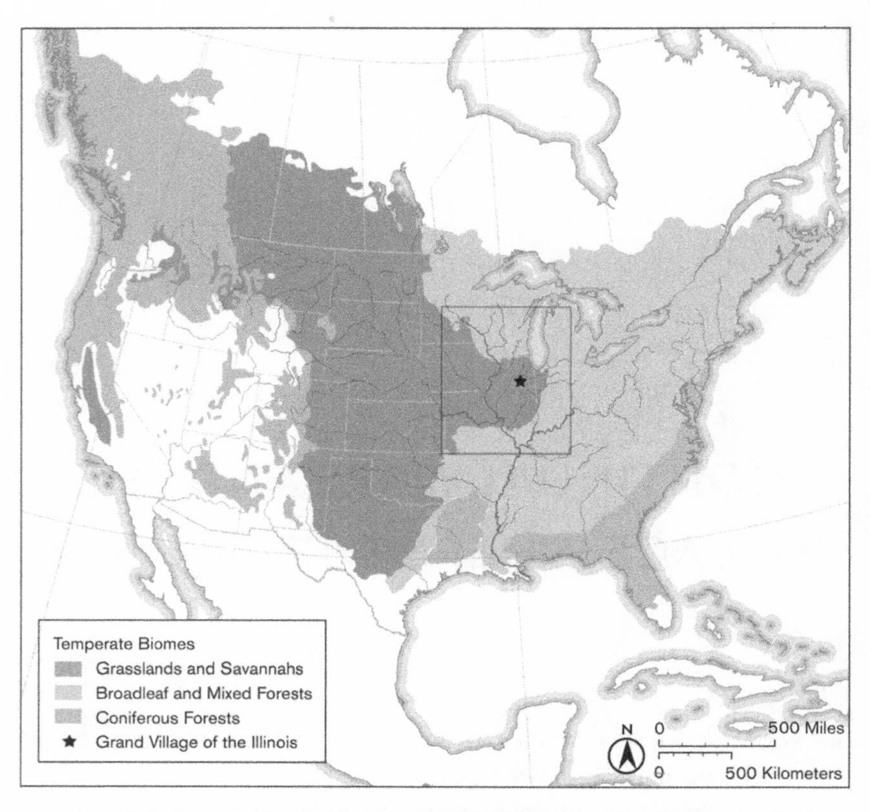

Biomes, North America. A major ecological dichotomy formerly characterized the central part of the North American continent, now dominated by monocultural industrial farming. Map by Ben Pease.

modern-day Indianapolis. Near the center of the zone and embodying all of its key features—multiple forest types, prairies, wetlands, and savannahs—was the river valley and drainage basin of the Illinois River, lying just on the other side of the Great Lakes watershed and draining approximately 28,000 square miles into the Mississippi Valley. And although this area has lately (as in the last hundred years) become one of the most monocultural landscapes on the planet, its previous incarnations were the opposite of homogeneous. Instead, the key characteristics of this region—what this book calls the "eastern prairie ecotone"—were diversity, variety, and edginess.[1]

I began writing this book because I wanted to know what this stark geographical reality—this *edge in the center*—had to do with human history. How did this huge ecological transition interact with human events here?

Given how we often assume the midcontinent's landscape and human history both to be monotonous and uninteresting, I wondered if we were missing something important about the region's past. I wondered whether there wasn't a hidden story to tell about the period in this land and region's history, before it became the monocultural industrial farming countryside we know today.

Answering these questions means confronting a collective amnesia about this specific landscape, so often dismissed and ignored. It is hard for us today even to imagine the environment of the eastern prairie region in premodern times. Since the arrival of European colonists, the region is probably one of the most radically and thoroughly changed environments on the planet.[2] Riding through Illinois on a bus sometime in the mid-twentieth century, ecologist and wilderness advocate Aldo Leopold lamented how invisible and unknown the ecological history of the prairie region was, even to contemporary farmers and residents, sadly observing that in a landscape defined by industrial farming, "no one sees" the complex past of the prairie region.[3] Surely his observation has only grown truer three generations later. Traveling across this countryside today on county roads or interstate highways, drivers see even less evidence of the premodern land cover than they would have in the 1940s, and almost no signs by which to measure the thoroughness of the land's alterations. If many accept the modern landscape of corn and soybeans as more or less the "natural" shape of a rural environment, or perhaps "second nature," as historian William Cronon once put it, that is understandable.[4] Vanishingly few relics remain of the previous vegetation. Rivers in this region have been so radically altered that presently it is impossible to sense the ancient hydrology of the region. And while major landscape transformations in other regions have left clues about their processes right there on the land—big dams, mines, irrigation and other infrastructure—the modern transformations of this eastern "corn belt" have left behind very little evidence of how they were accomplished. Much of what changed the environment in the past century is invisible. It is either gone, illegible, or, like the thousands of miles of underground drainage "tiles" that may constitute the most profound landscape alteration in many parts of this region, literally buried in the ground.[5]

As Leopold lamented in the 1940s, however, there's another reason why we have so much trouble seeing the premodern landscape of this region, and it has to do with our own myopia, so to speak. There's no *story* in the popular imagination about the natural history of this region, whose ancient land cover has been so thoroughly transformed.[6] There's not a good sense of what it was nor how it was shaped, let alone the very important human history that went along with it. In this, one of the most privatized parts of the United States, there are few public lands and no national parks preserving the pre-Columbian conditions in the region let alone much of a sense that that might have been a good thing to do.[7] And if it is true, as several historians have recently argued, that human history in the Midwestern region broadly is more thoroughly forgotten or simplified than it is in other regions of the United States, it seems reasonable to suggest that the failure to see the environment and the failure to see the human history of the eastern prairies are indeed connected in a circular way.[8] "Illinois has no genesis, no history," Leopold lamented, "no shoals or deeps, no tides of life and death."[9]

Living in east central Illinois, in the heart of this former region, I think about Leopold's observations all the time. As a historian of early America I am keenly aware that my home region's human stories, no less than its ecology, have often been purposefully simplified and sometimes expunged. Little celebrated among Indigenous peoples in the national narrative, the many groups who dominated this place in the premodern period—the Meskwakis (or Renards, or Foxes, as the French often called them), the Myaamias, and especially the Illinois—were for a time surely the most powerful Native people of the interior, perhaps on the continent. A settler colonial culture in thrall to its innocence and narratives of what Jean O'Brien calls "firsting" and "lasting" both passively forgot and actively erased the stories of these Indigenous peoples from the land after a long process of dispossession and removal in the Midwest.[10] Despite this, not only do the descendants of these prairie people still survive and claim sovereignty in this land, but their long histories can be recovered and revalued in regional memory. Focusing on the prairie landscape and its entanglement with these peoples reveals a fascinatingly complex story of Indigenous power, a true odyssey of the tallgrass. This is not just an opportunity for better regional consciousness, it is a matter of getting our history right. The widespread and consistent failure today to understand the nature of this ecotone zone, or the nature of its past, has caused us to miss one of the most important chapters in the history of early America. Here is one story to help us appreciate the significance of this forgotten and singular region, a case study in how it shaped and was shaped by dramatic human events.

◆ ◆ ◆

To read eyewitness accounts of European travelers to the midcontinent of North America in the seventeenth and eighteenth centuries is to read a great deal about violence. Crossing beyond the Great Lakes watershed and into the Mississippi Valley into what Jacques Marquette in 1673 called "strange lands," Europeans constantly remarked about war-torn and ravaged Indigenous populations.[11] These eyewitnesses narrated the stories of violent attacks, reported on the devastating aftermath of raids, and tried to explain—in their often biased ways—what was causing the carnage. Whether framing Indigenous people primarily as agents of violent encounters or as victims, many observed the key fact: warfare was shaping lives profoundly. Above all, these eyewitnesses paid attention to a grim bottom line: through the maelstrom of violence and warfare—*continual wars,* as traveler Jean-Benard Bossu described the situation after voyaging through the region in the 1750s—Indian people like the Illinois, the Meskwakis, and the Myaamias were declining and decreasing in numbers.[12] Despite living far away from any major center of colonization, in what we would have to call a "Native ground"—borrowing a term by one influential scholar of early America—the region of the modern Midwest, and especially this eastern prairie ecotone, was as war-torn perhaps as any other place on the continent.[13] Dramatically, it even witnessed one of the most severe Indigenous conflicts of early American history, the Fox Wars, which lasted from 1712 through their climax in 1730 in a prairie battle near modern-day Arrowsmith, Illinois, costing the lives of hundreds. Often overshadowed and misunderstood in regional memory, these Fox Wars, and the larger patterns of violence to which they belong, are among the central reasons, together with disease and other factors, why Indigenous groups like the Illinois so quickly lost their status as "the most numerous," as Bossu once put it, of interior America's people.

While frequently describing them, however, French eyewitnesses had few tools to explain these patterns of Indigenous violence. On the one hand, they understood the outlines of practices like mourning war, the distinctive patterns of warfare fueled by cultural imperatives in Algonquian societies for families to replace lost kinsmen in battle. This phenomenon, the French understood, helped produce especially intense bursts of warfare in key moments. But these bursts were a near-universal feature of military culture among Great Lakes Indians, and did not by themselves explain why violence by and against Indians like the Illinois, the Meskwakis, and their neighbors was so dramatic or long-lasting.[14] In many cases French eyewitnesses chalked up the violence they witnessed—or, more often, heard about—to some supposedly innate qualities of certain Indians themselves, casting them as essentially warlike and aggressive or, in the cases when Indians suffered conspicuous *losses*, "not at all warlike" and thus primitive and deficient in battlefield ardor.[15] Of course they remarked on the fur trade rivalries among groups in the region, and they surely paid attention to particular enmities, if only so that they could exploit these divisions for their own purposes or, conversely, try to reconcile them to facilitate colonial agendas. But rarely did French eyewitnesses try to historicize or fully understand the motives behind these rivalries or the specific reasons that explain why violence became so intense and pervasive. Indigenous warfare, they often concluded, was just the "general custom of the country," driven by what was effectively an irrational bloodlust rooted in vague and ancient enmity.[16]

For their part, historians have done better, but not enough. To be sure, historians of early America have done much to relativize warfare and place it in a broad historical context, effectively and totally demolishing the notion that Indians were simply or in some essential way either warlike or primitive.[17] But, while echoing the lessons of this larger historiography, historians of the midwestern region broadly have focused more attention on the mechanics or effects of the intense violence of the midcontinent, rather than understanding its roots. When they have looked to causes, they have stressed discontinuity, positing that colonialism overwhelmed "older patterns and older routines" for Indigenous peoples.[18] They have told a story which, while featuring the Indigenous logic of mourning war, nevertheless most certainly privileges outside forces—colonialism, exogenous disease, and invasion—as the most important causal forces *acting on* an Indigenous Midwest. In some cases these stories of invasion have even reified the French notion that Indigenous people of the prairies were weak, dependent, or desperate.[19] Even some of our best historical accounts have followed the logic of French sources and implied that much of the violence of the Indigenous Midwest was merely reactive, a matter of a mostly unchanging Indigenous world buffeted by the rupture of contact and market-making.[20]

This emphasis on colonialism and Indian *reaction*, however, is in tension with one of the great lessons of recent early American historiography. In the past two generations historians have reframed the story of early America fundamentally by putting Indigenous peoples at the center. On one level this has required simply changing the perspective—*facing east*, as one historian put it—and reconsidering familiar events from an Indigenous vantage point.[21] But a more profound innovation in much of this scholarship has been to decenter colonization itself to recall the many moments and places in which colonists were not nearly the singular driver of change that generations of historical accounts have attested. To the contrary, in many places and for long stretches of time it was Indigenous agency and Indigenous logic that shaped history.[22]

In the most far-reaching examples of this scholarship we learn about Indigenous empires and Indigenous peoples who were not victimized by colonialism but who themselves practiced "reversed colonialism" on their neighbors and on Europeans.[23] We learn, as mentioned, that many places in North America remained solidly Native ground, affected by the arrival of Europeans but mostly shaped by ongoing Indigenous agendas. Thanks to this scholarship we understand Indigenous worlds and "interior zones" as places that reached out and incorporated colonial processes as much as they were enveloped *by* those processes.[24] We understand Indigenous systems of slavery and the ways in which Europeans were often forced to abide by Indigenous expectations of violence, not vice versa.[25] We learn about individual groups who rose to become "masters of empire," flipping the script and recasting the taken-for-granted power dynamics of colonial encounter.[26] Through it all we have not only learned to distrust the usual stories of colonial American history, with their just-so air of inevitability and teleology, but we have also learned to remember the simple context that makes those narratives so dubious. The colonists may have wanted to make the Indians "dependent" on them, to totally reshape their world and actions, as the explorer Robert La Salle put it in the 1680s.[27] But then LaSalle wrote this as he traveled with a small party of forty Frenchmen into an Illinois River Valley occupied by thousands of Indigenous people. However loudly they speak in voluminous archives, the voices of the European colonists were weak on the ground in many corners of the Indigenous world of the seventeenth and eighteenth centuries, including especially in the midcontinent.

The lessons of this scholarship, of two generations of asserting Indian agency, leave us with both an interpretive dilemma and an opportunity. For decentering the colonists requires us now to explain in Indigenous terms the origins and logic of the maelstrom of violence of the contact era. Take, for instance, the climactic and horrific battle of the Fox Wars in 1730, in which at least five hundred Meskwakis were killed and more than three hundred more were captured by an allied force of Illinois, Potawatomis, Sauks, and Miamis, along with a smaller French force.[28] In many typical accounts events like these are glimpsed through the window of French sources, and particularly discussions by French planners who arrogantly stated their intention to "destroy this audacious and rebellious tribe."[29] But while French people may have *wanted* to exterminate what they viewed as "a cunning and malignant" group, it was Illinois-speaking people who actually had the power to nearly realize the vision.[30] Surely we must be extra careful here. We know colonists purposefully created stark images of Indian violence as a discursive tool to solidify settler identity and justify colonialism; settler-descended scholars like myself have been raised inside such nationalist mythologies.[31] Yet, as Karl Jacoby and others have argued, it is precisely because of such influential discourses of Indian brutality that it is necessary to look unflinchingly at the violence itself and contextualize it in its fully comprehensible logic. Only through such an approach, one that foregrounds Indigenous motives, unplanned contingencies, and the combination of action and reaction that many violent acts entailed, can we move to see a more complex and human story about the violence at the center of early America and its hard outcomes.[32] This means not losing sight of settler violence perpetrated on Indians, nor the entanglements of Indigenous violence with colonial forces. But it also means taking advantage of Native-centered approaches, to reveal violence as a constellation of Indigenous power, strategy, and division, some of which was only distantly related to colonialism. It means keeping the important particularity of early America in mind and understanding its connection to later processes of settler colonialism, but also understanding the very different dynamics of power in a period long before settlers arrived in any large numbers.[33] Such approaches help move us past the simplistic historical narratives that frame all Indigenous actions in terms of colonialism and as *reactions* to exogenous factors or symptoms of supposed dependency, while also giving us new ways to understand the burdens of that colonial past.[34]

Inspired by recent writings about early America, this book takes a new approach to understanding the Indigenous history of the midcontinent and its climactic conflict, the Fox Wars. While not denying the important role of colonialism in shaping events even in such a remote region as the far interior, I argue that extraordinary violence in the midcontinent needs to be understood as well in terms of what we can infer about Indigenous motives and in the ongoing context of Indigenous history, including especially in the period well before colonists arrived.[35] Tracing one of the most interesting examples of the rise and fall of Indigenous power in early America, I argue that violence was essential to a longer history of the Illinois, the Meskwakis, the Myaamias, and other neighbors, and not the incidental or solely exogenous result of colonialism. At the same time as violence was essential, however, it should not be *essentialized*.[36] Above all, I argue that the maelstrom of violence that defined the history of midcontinental Indigenous people was not an abstract or general phenomenon of culture, but rather a historical one that was created *in time* and, most important, *in place*.[37]

Bringing together insights from both the new Native-centered history with environmental history, the premise of this book is that we can best understand what was happening in the Indigenous history of the eastern prairies—particularly its violence—only when we pay close attention to the unique dynamics of the special biophysical region in which it happened. Challenging our usual tendency to think of human agency—whether Indigenous or colonial—as independent of nonhuman realities, this study narrates a history shaped by the entanglement of people and particular "natural" forces. It argues that we need to understand human events of this period as growing out of a unique bioregion, a geography of division, in one of the most important and dynamic borderlands of early America.[38]

Although not frequently recognized as a major setting for early American history, this "edge" at the center of the continent was a special place and the heart of a massive and disappeared landscape: the tallgrass prairie. Characterized in 1935 by pioneering ecologist Edgar Transeau as "the prairie peninsula," here was a huge region that stretched for hundreds of miles from east to west, with its eastern edge usually running approximately through modern-day Minnesota, Wisconsin, Illinois, and Missouri. For reasons of scale and focus, this book concentrates primarily on one important part of this larger zone, or what might be considered the "point" of Transeau's peninsula. Bounded by the Wisconsin River on the north and the Wabash and Ohio Rivers on the south, and encompassing the entire Illinois River Valley, this part of the tallgrass transition zone witnessed a particularly momentous series of events in both the human and nonhuman realms over a *longue durée*. It was a kind of heartland in the midst of the larger and more generalized ecological transition that covered the midcontinent.

This part of the tallgrass peninsula was also a significant cultural borderland for Indigenous North America in the contact era. The entire prairie-woodlands transition zone roughly approximated a division between broadly contrasting Indigenous culture groups who had dominated it in the wake of the Mississippian chiefdom at Cahokia. On the eastern side of the divide, Algonquian- and Iroquoian-speaking groups and polities dominated in the Great Lakes and Ohio River Valley regions. On the western side, Siouan- and Caddoan-speaking peoples settled mostly in the river valleys of the enormous

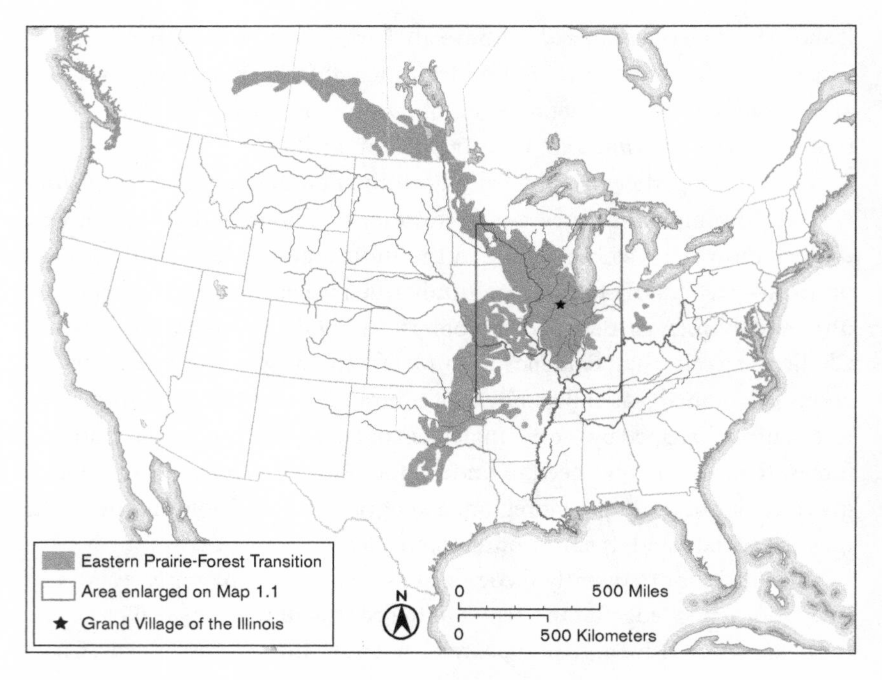

Shoreline of Grass. The "prairie peninsula," named by ecologist Edgar Transeau in 1935. Before it became the monocultural landscape we know today, this region was the edgy tallgrass prairie mosaic, one of the great ecological transition zones of the North American continent. The Illinois Valley formed a kind of "point" in the larger peninsula. Map by Ben Pease.

grasslands west. Surely these groups were not internally unified in any sense, politically or culturally, nor was there any kind of boundary preventing interaction among peoples across this broad division. To the contrary, interaction across the midcontinent was constant and ongoing over generations, and many traditions and cultural concepts spanned the whole zone.[39] Just like the ecological transition, this cultural edge was blurry and the categories of "plains" and "woodlands" peoples were not perfectly exclusive. What did seem to *divide* them somewhat consistently along the broad and general transition zone, however, were kinship lineages and group histories, a fact which shaped and limited certain kinds of interaction and made the region's peoples distinct in ways that went beyond obvious differences in lifeway tied to environment. In any event, for reasons that will soon be made clear, the Illinois Valley was a crucially important part of this cultural borderland as well.

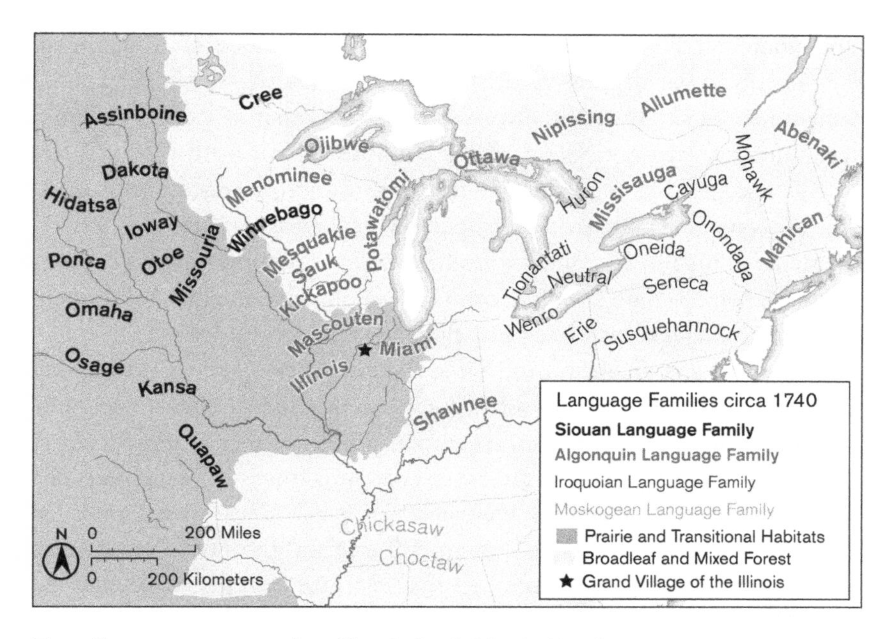

The tallgrass ecotone as cultural borderland. Map by Ben Pease.

Historians have largely ignored this place as a central location in American Indigenous history, but they have been wrong to do so. The contention of this book is that this region helped shape an important history, and for specific reasons. The central intuition of this book is that for several generations before and after the period we generally think of as the contact era, the diversity of the transition zone—both in ecological and cultural terms—provided opportunities for certain kinds of subsistence as well as strategic endeavors for the people who inhabited it and could exploit them. Although French eyewitnesses could easily see the distinctiveness of the environment and the lifeways pursued by prairie inhabitants, what they often missed were the ways in which these lifeways strategically maximized certain special characteristics of the zone.

In making this argument, this book invokes a concept from ecology: the "edge effect." Writing in his classic text, *Game Management*, ecologist and early environmentalist Aldo Leopold first identified what he called the "law of interspersion," or the idea that areas of overlap among different plant communities—such as, say, forests and wetlands—are the place where many animals can thrive by taking advantage of multiple opportunities. "Game is a phenomenon of edges," Leopold wrote, implying that animals maximize their energy in the locations where they can exploit multiple ecologies at the same time. The key is "*simultaneous access* to more than one environmental type" and the "*greater richness*" of opportunities in these settings.[40]

At the center of Leopold's analytic, then, is another concept from ecology: the ecotone. Discussed first by pioneering American ecologist Frederic Clements in his influential early studies of plant communities, an ecotone is a special place where one plant community shades into another and where different ecologies come together along a stress line or tension zone. Derived from the Greek roots meaning "two houses in tension," the word ecotone refers to places where communities come together in a complex configuration. This concept speaks to the history of the midcontinent and particularly the heartland of the eastern prairies. An ecotone on many different registers, including quite literally in the biological sense as well as the metaphorical sense in terms of contrasting human cultures and lifeways, the tallgrass prairie region in the premodern period gave its inhabitants many material opportunities, creating edge effects that shaped history.[41]

If historians have largely overlooked how divisions in this region created special opportunities for its Indigenous inhabitants, we have completely missed what was even more important: the *historical dimension* of these opportunities. For if the region was an important borderland, it was also dynamic, changing, and in the process of a crucial species shift when the contact era opened. Several generations before colonists showed up in the Midwest, and during a major reshuffling of Indigenous America in the wake of the Mississippian period, climate change and other factors contributed to the arrival of an all-important actor on the midwestern historical stage: the bison.[42] As archaeologists have recently confirmed, the large and "innumerable" herds of bison grazing on the tallgrass prairies at the time of contact probably represented a historical novelty, a kind of ecological release, although a complex one. They were perhaps the first large herds of bison grazing in these far eastern prairies during the whole Holocene epoch. Although too often ignored, these animals shaped Indigenous cultures of the eastern prairies in distinct ways and led to the rise of new bison cultures in a specific moment of change. Here was one important part of the region's edge effect, but it was not some timeless regional reality. It was historical. It was *new*.

The significance of this historical context should not be understated. For if the environment was changing, it was probably joined to an important change in the human population of the region in the same period, just before colonization began. In particular, all the main protagonists of this book—the Illinois, the Meskwakis, and their sometimes neighbors, like the Myaamias—migrated or sojourned into the tallgrass prairies from the east, beginning in the 1600s, likely attracted by this species shift. They arrived in a region where previous occupants had begun a bison-based lifeway, developing a special economy as *pedestrian* bison hunters. In arriving here the newcomers changed their lives, adapting to the new environment. They took advantage of the opportunity that the bison afforded while remaining in contact with old networks. The species shift and the resulting edge effect shaped a brand new human history as these people rose to power.

The basis of that new power, of course, was energy.[43] While diverse and edgy, the most notable vegetation cover of the Illinois Valley and surrounding region was a grassland full of plants capturing solar energy and converting it through photosynthesis into biomass.[44] Importantly, there was as much or perhaps more primary productivity in this grassland landscape—interspersed with wetlands and wet prairies—relative to forests in the east. (This is an intuitive fact for anyone who has visited the corn belt in June; the midcontinental region is today one of the most intense regions of photosynthesis on the continent, and it was then too.) But where corn belt farmers today plant annual grasses such as maize to tap the sun's energy directly, the energy in C4 perennial grasses, forbs, and herbaceous plants that dominated the eastern tallgrass prairies before the invention of the steel plow were largely locked up—because inedible—to humans. Although Indigenous peoples of the midcontinent had been farming for millennia, and growing corn in particular since the early 1000s, they really could not farm the prairie uplands owing to the dense plant roots underneath those grasslands. In the prairies, people needed an intermediary. With bison, which could graze on the grass, humans had their partner, their intermediary, their reservoir of accessible biomass.[45] Here was a major opportunity. The people of the tallgrass prairies had simultaneous access to their original home in the woodlands, which they could still access via broad river valleys along the Illinois and the Wisconsin, and they also had this new bison-based energy economy that their eastern neighbors lacked. This opportunity shaped them.

This book narrates how the interrelationship of people, bison, and the larger environment drove prosperity for these specific cultures. But this was not a case of simple determinism. Instead, people entangled themselves with new ecological realities in ways that challenge the usual distinctions between the "natural" and the "human." Foregrounding the relationality of humans and nonhumans in the tallgrass, particularly through distinctive kinds of *work*, it's evident that bison, fire, grass, and human labor all shaped each other in what might be considered an assemblage, or *collectif*.[46] Less familiar than certain other human-animal cultures like cattle ranching or sheep herding, not to mention the famous equestrian bison cultures of the North American plains, the unique pedestrian bison hunters of the tallgrass prairies were no less shaped by their material relationship to a special animal in a special place.[47] Some ways this was true are obvious, of course, such as a cyclical and seasonal lifeway and a distinctive material culture. But in demanding certain kinds of work, pedestrian bison hunting shaped more subtle biophysical and cultural aspects of life. It enabled and required certain kinds of social organization, including large (and linked) villages. It underlay certain kinds of cooperation, including distinctive complementary gender relations. And, of course, it forged a special worldview of reciprocity, mutuality, and community with a particular nonhuman entity and environment.[48] Historians have overlooked the significance of these realities, which were shaped by chasing bison on foot, but we should not. Part of the agenda of this book is to bring to life the unique pedestrian bison culture that was created by these easternmost "bison people" in American history.

While the opportunity of bison hunting shaped their lives, however, it is also important to remember that these people were not alone. The Meskwakis and the Illinois accompanied each other in their respective migrations to the prairie edge, with the Miamis not far behind. The Potawatomis rose in the region during this period as well, and the Shawnees and the Ottawas sojourned here too. Meanwhile, already present in the eastern prairie region were groups like the Ho-Chunks and their Siouan-speaking neighbors to the northwest. Caddoan speakers like the Pawnees occupied the prairies to the southwest. The Chickasaws occupied prairie edges in the southeast. And, perhaps most notably, Iroquois and Sioux neighbors—the Lakotas and the Dakotas—were present in the region beginning in the post-Cahokia period and their presence just on the margins of the prairie ecotone was hugely significant.

The important fact is this: many different people were simultaneously making their lives, at least to some extent, on this eastern prairie borderland. And, particularly in the case of the Illinois and Meskwakis, these people were *competing*. This made history too. Expressing itself often through the distinctively old but now evolving practice of captive-raiding, this competition entangled old imperatives of kinship replacement—the "mourning war"—together with novel conditions of the pedestrian bison-hunting lifeway—such as large village—to produce an increasingly violent history. Captives became newly important, both as symbols of alliance among these large and diverse villages as well as laborers in the bison economies that underwrote them. As bison hunting subtly shaped aspects of social life, so did it subtly but inexorably shape tensions among people living in this special geography, divided so importantly among newcomers and longtime residents, kinsmen and strangers.

When colonization began, all of this history was ongoing and in process. And only *then* came the new forces that colonization set loose in the region. To be sure, the normal mode of understanding the Indigenous history of early America is to assume that colonialism changed everything, and that is surely somewhat true. Yet colonialism produced its specific effects only by interacting with these equally specific and ongoing processes and the momentum of ongoing trajectories. To understand the dynamic results of colonialism we need to start with this long runway of Indigenous process, and we especially need to understand the entangled role of humans and nonhumans in shaping it. For the upshot is this: if bison hunting evolved with specific social dynamics within Indigenous communities in a co-constituted fashion, this same phenomenon of co-constitution determined how new colonial forces—new diseases, the fur trade, guns, and the like—shaped and were shaped by the undercurrents of a powerful, changing Indigenous world. Indeed, colonialism's profound effects in the prairie region can be understood only in the context of this long Indigenous trajectory and its entanglement with a special nonhuman environment. It was this trajectory that shaped *how colonialism came to matter.*[49]

Nowhere is this truer than when it comes to the headline issue of violence. When in 1750 the traveler Jean-Benard Bossu looked back on the "continual wars" that shaped the history of the prairie people like the Illinois, he was probably referring to events that he himself knew about, specific events punctuating the region in the 1710s and climaxing in the Fox Wars in 1730. But Bossu's word "continual" was also apt for describing a deeper past than he probably understood. Far from becoming irrelevant at the start of colonialism, that long history was fundamentally shaped by how the new forces introduced by French colonists played out during these violent moments. In no way was the arrival of the French just one more in a long set of changes, or the arrival of just another newcomer like all the others. But if the French—and their diseases and marketplaces—were different, it was because the significant changes they brought plugged into existing currents, in effect simultaneously charging them and drawing intensity from them. In substantial ways the arrival of colonial forces interacted with ongoing cycles to produce the kind of violence that was so notable to French eyewitnesses and so consequential for Indian history.

All of this is especially true for the Illinois and their participation in the Fox Wars. It was the Illinois, this book argues, who frequently were the most important perpetrators of that conflict. It was the Illinois who most strategically adapted the logics of captive-raiding to pursue alliances and build power in their diverse region. It was the Illinois who most purposefully committed to—and then doubled down on—the special opportunities of bison hunting, amassing power while also exposing themselves to an unstable ecology and volatile social arrangements. To understand their role and their logic we need to understand these ongoing Indigenous histories and ecological contexts. Historians have mostly understood the Fox Wars by centering the Meskwakis and their complex diplomatic strategy, and—especially—the French diplomats, whose efforts to control them are so visible in the colonial archive. But the Meskwakis' complex history in and alienation from the Great Lakes alliance was rooted most importantly in themes often obscured in that archive: the long-standing animosities of their special regional history and the strategic actions and decisions by themselves and the Illinois. In other words, the Fox Wars originated in the special vengeance that developed between groups whose strategies, including material strategies, clashed so strongly—because they were both so similar and so precarious—at a moment of change.

In the end this book offers a new way of understanding the significance of this distinctive place and its people in early America. Rather than focusing on an individual tribe, it centers a dynamic zone. Rather than starting with the rupture of colonialism, it concentrates on events that shaped Indigenous motivations well before first contact and continued to do so profoundly right through the mid-eighteenth century. Rather than a simple story of natives and newcomers, this book examines processes of encounter and contestation among Indian peoples themselves, a kind of Indigenous *métissage* and culture-creation in the generations before contact. And rather than foregrounding French agendas to explain the momentous violence in the early Midwest or the invasion of the region by outsiders, this book explores the long Indigenous and material roots of transformational events like the Fox Wars.

All of this attention to material context and the trajectories it shaped makes our history a much more complex and complicated story, however. At a fundamental level the experiment of decentering that this book attempts results not just in a new story about the relationship of Indigenous history and colonialism and the question of what weight to assign to different human agencies in creating historical outcomes. More fundamentally, this approach makes us decenter humans altogether, to see instead the rich entanglement of people and nonhumans as always deeply shaping how history plays out. While the primary purpose of this narrative is to tell a better story of Indigenous power in a special and ignored region of early America, there's no doubt that a side effect of this narrative is to trouble our naive certainty about more abstract questions: Who in the past was actor, and who was *acted on*; What was cause, and what was effect? For if a bison economy itself helped shape the specific severity of disease epidemics by concentrating unusually large numbers of people in a wintertime hunting camp, which in turn accelerated demand for captives and therefore violent raids, who or what is the "agent" of this history? Simple stories that center only colonialism, or that cast Indigenous people as only reactive, ignore the long contexts and histories of place and need to be cast aside. So, too, do simple ontological certainties about "natural" forces, "human" decisions, and the worlds in which they operate.[50]

Above all, we need to add new complexity to how we understand Indigenous peoples in the contact period. The Indigenous history of the Midwest in the colonial period broadly has been told in terms of reaction, with Indian people responding or "adapting" to colonialism, shaped by declension, and reduced to dependency. Better narratives have emphasized Indigenous actions, but in terms that still too often assume a basic teleology of colonialism and invasion overwhelming a mostly stable Indigenous world at the all-important rupture of contact. Without underestimating that rupture, this book offers a new narrative, one that uses untapped archives to reveal a prior Indigenous world in motion, a dynamic and ongoing world of change and action featuring people taking advantage of new and unique opportunities in one of the most important transition zones of North American culture and ecology. Foregrounding how these ongoing processes fundamentally shaped how colonial forces played out, my aim is to move past clichés like adaptation and dependence to tell a more nuanced story, one that helps reveal the power of Indigenous people and their odyssey in this special region, the edge in the center of the North American interior.

◆ ◆ ◆

In a series of letters describing the Indigenous peoples of the Great Lakes and prairies in the early 1700s, New France's intendant, Denis Antoine Raudot, wrote a notable passage about the conflicts that would soon become known as the Fox Wars, rendering a stunning and stark observation about the war's main protagonists, the Foxes, or Meskwakis. Describing them (as well as their oft-allies, the Kickapoos and the Ho-Chunks), Raudot wrote that their violent history could be laid to a simple cause: the people themselves. To understand the causes of the Fox Wars one just needed to understand these peoples' essential nature: "One could not treat otherwise the last three nations I have just mentioned than to say that they are devils on earth, they have nothing human but the shape . . . One can say of them that they have all the bad qualities of the other nations without having a single one of their good ones."[51]

Stated strongly, Raudot's thinking here is not so unusual. It is rather a good example of how the French interpreted the Indian conflicts of the eighteenth century in terms of a logic of essentialism. There was no sense to Indian violence, the French thought, just a basic maliciousness and wantonness that could never be truly comprehended. If Indians fought each other and the colonists to their own devastation—while ignoring French attempts at mediation and continuing to resist long after other Indigenous groups had accepted French peacemaking—it was just because of their "bad qualities."

Historians have done a better job of explaining the violence in early American Indian Country and moved well beyond Raudot's essentialism to come to a richer understanding. Although relying on the sources left to us by those same French observers, they have explored the logic of mourning war and a so-called engine of destruction to appreciate Indian dilemmas. They have highlighted Indigenous resistance and long-term strategies. In so doing they have begun to make the violence of early America comprehensible in terms of Indigenous agency and the struggle to survive. They have started to make Indians the center of the story and dismissed essentializing narratives.

Nonetheless, our narratives of early American history still too often ignore certain dynamics that were central to Indian actions and the hidden logics that drove them. In the case of midwestern groups who occupied the eastern prairies, our narratives foreground invasion and cast Indians as reactive. We take for granted the agency of exogenous factors like *guns, germs, and steel* and all too often miss how those factors were mediated and given causal efficacy by equally important local realities. Our narratives usually start slightly before contact but miss the long trajectories and deep-time perspective that shaped the context in which colonialism happened. While advancing our understanding far beyond the old essentialist narratives, historians have not gone far enough in explaining the dynamics of place that often drove Indigenous actions, particularly in the prairie ecotone. So far our understandings may be said to have stopped short.

A small but telling example of how historians have stopped short in understanding the Indigenous history of the midcontinent might be found quite literally in their use of Raudot's observation quoted above. The evocative line "devils on earth" appears in several modern histories, excerpted like this to illustrate the way the French understood and dismissed their enemies. But it wasn't until I went back to the original document that I learned that Raudot's observation about the Meskwakis and their allies does not end with the sentiment about "bad qualities." In the very next sentence Raudot continues and elaborates his idea. He hints at a more complex and curious understanding of the Meskwakis' war-torn history. As he put it, "The *place where they live* is well situated for living and *seems to make them more ferocious and more insolent*" (emphasis added).[52]

To be sure, the line just hangs there, and Raudot quickly ends his letter thereafter. And given the prejudicial language Raudot deploys here, it is surely easy to dismiss the sentence out of hand as meaningless and bigoted cant. But linger on the observation for a minute and try to take it seriously. It does not require too much critical insight to suggest that what Raudot saw as "ferocious" and "insolent" behavior might be read in a more neutral way as the Meskwakis' and their neighbors' strategies for dealing with colonialism and their rejection of it. And once we recast the sentence with that in mind, supplying even a tiny bit more historical empathy than Raudot could muster, the rest of the formulation takes center stage: the idea that the Meskwakis' distinctively violent reaction to colonial agendas, their unwillingness to defer to or trust a French alliance system, or their general hostility toward their neighbors might stem—somehow—from the "place where they live." It is a short line, but more than intriguing. What did Raudot mean?

Well, let's just be clear: *the place where they lived* was the eastern tallgrass prairie region. It was a biome-scale transition, an edge in the center of the continent contested by newcomers and ancient inhabitants alike. Historians have hardly paid it and its special geography—not to mention its fascinating history—any attention at all. What was Raudot getting at? What might we have missed about *the place where they lived*? To find out, let's start in the *obvious* place: in the grass.

1

SHORELINE OF GRASS

NOBODY TODAY ASSUMES THAT THE PRESENT MIDWESTERN LANDSCAPE cover of industrial corn and soybean agriculture defined the ecology of this region prior to one hundred years ago, let alone long before that. Yet we do frequently assume a certain continuity between the environment of the Midwest's present and that of its distant past, including the eastern prairie ecotone. Specifically, many people project the current-day environment's ecological simplicity and monotony into history, sometimes even imagining the preindustrial corn belt as perennially homogeneous and perhaps somehow *monocultural.* The best evidence of this habit of mind is in the metaphor that we use to describe the ancient landscape cover of the region: the ubiquitous notion of a vast and unbroken "sea of grass." This metaphor, which has been in broad use in discourses about the eastern prairie landscape since the eighteenth century, invites us to imagine that even though the prairies that thrived here before industrial farming were surely bustling with huge numbers of species (with few people appreciating just how many), this likeness to an ocean stemmed from the basic invariance, the uniformity, and a fundamental lack of diversity that the prairies shared with the modern corn belt landscape.[1]

We need a complete reenvisioning of the prairie's past. Despite the power of the image of an ocean of grass, it is nevertheless a poor representation of what existed in this easternmost grassland zone at any time since the beginning of the Holocene, or the end of the most recent ice age. Rather than an unbroken homogeneous expanse, here was a patchy "ragged edge," to adapt F. Scott Fitzgerald's famous descriptor to an ecological context. Indeed, this region was less a *sea* than a *shoreline*, a threshold that brought together diverse vegetation communities and ecozones into rich edge formations. Like a shoreline, this landscape featured not just the easternmost part of North America's grassland "sea," but also "estuaries" of lowland and riverine forests of at least three different types as well as "islands" of upland forests and "reefs" of oak savannahs and groves. Perhaps even more difficult to imagine, the landscape featured quite literal shorelines: massive wetlands and huge stream-and-river systems like the Illinois, with its enormous floodplain and backwater lakes. It was defined not by monotony but by great diversity, by edginess. Moreover, the very flatness of the midwestern landscape—the characteristic that makes it so fundamentally uniform and homogenous in our imaginations—is part of what made it so diverse and edgy.

There is another equally important dimension to the debunking we need to do, however. In addition to viewing the presettlement landscape as a monotonous sea of grass, we too often think of it as a timeless wilderness shaped by unchanging "natural legacies" in the far distant past.[2] This is perhaps an even bigger blunder. A better understanding comes when we push the shoreline metaphor a little further to recognize how the edgy vegetation of the midcontinent was not only clashing; it also was constantly moving and flowing. Like a beach, it was a blurred transition in which the edges of different vegetation communities shifted against each other, extending and receding in slow motion like waves—or perhaps tides—over a flat surface. Rather than a timeless wilderness, the prairie-forest transition was the product of distinct change, much of it relatively recent in the perspective of evolution or deep time. Moreover, while much of this change to the landscape happened without people, as, for instance, when grassland formations pushed east into the Great Lakes or forest edges pushed back due to climate change, for several thousands of years the landscape was also in certain ways—and at certain scales—a human creation, the result of a complex collaboration between human and nonhuman forces. From the moment that people began to wield fire to construct and shape the region, humans have been the most important force in the landscape's motion and variation, and human influence here was sometimes so thorough as to deserve the classification of a kind of domestication.[3]

Generalizations about the shaping of the ecotone landscape need elaboration, however, especially if we want to create a better narrative for the environmental history of this region. Because while people were crucial shapers of the landscape in many places in the premodern past, their role is one of the themes that has become rather abstract, and even clichéd, in our understanding of early American history. The shaping of the eastern prairie land was a series of events, not a principle. It occurred specifically, not generically.[4] And while the question of whether the prairie was an anthropogenic landscape has been explored many times, it has most frequently been framed simplistically, as a yes or a no, and out of context of specific events.[5] This is limiting. The crude question of whether people fully "created" the tallgrass prairie or not—especially when explored at an abstract level—precludes historians' ability to see the land as the product of *both* natural and human history. Moreover, the blunt framing of and partisanship on this issue has buried a more important prior story: the rise of distinctive characteristics in the region's makeup, which affected how and to what extent people could—and could not—control its especially sensitive ecological formations in the first place. Unburying this prior story means exploring several special dynamics of this "shoreline" region and their roots in deep time.

Ecologists and environmental historians commonly refer to precolonial Indigenous landscapes as *mosaics*. The metaphor comes out of the history of ecology. After years of considering plant communities as superorganisms, each with an inherent destiny of a single "climax," twentieth-century ecologists have revised their assumptions and posited instead that ecological assemblages are stochastic, nonlinear, and ever-changing. With the mosaic idea, ecologists moved away from a teleological notion that environments are stable and uniform and that their components acted together with some sense of equilibrium and harmony.[6]

The metaphor of the mosaic is useful not just because it suggests the diversity and variability of a landscape like the tallgrass, but also because it helps us envision its history and creation. Surely the end result of any mosaic rests crucially on the available tiles, or *tesserae*, that the artist can employ in her design, as well as on the relative "stickiness" of the relationship between the tiles and the underlying surface.[7] This is why the midcontinental grassland mosaic is so special. First, regarding the kinds of *tiles*, as it were, that were available in the region, the midwestern forest-prairie tension zone since the Holocene is a place where the full panoply of flora and fauna from two *biomes*, as well as from several other smaller ecozones, met and mingled. Regarding the *surface*, this region is distinguished as one of the flattest places

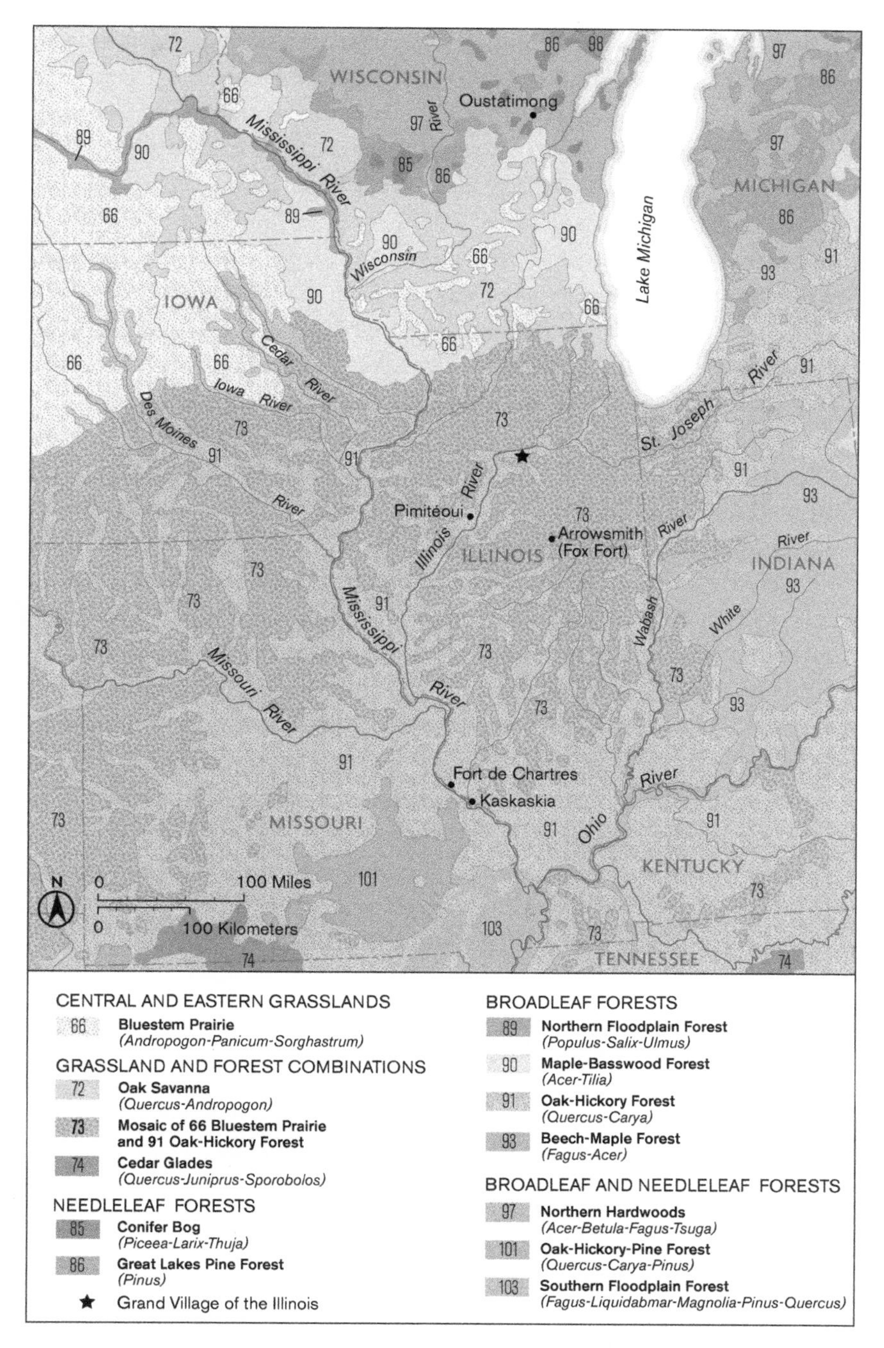

Eastern Prairie Mosaic. In his 1966 study of vegetation boundaries, "Potential Vegetation of the United States," A. W. Küchler defined the zone of the eastern prairie in terms of flux, a "mosaic of bluestem prairie and oak-hickory forest" whose boundaries were frequently in transition. Map by Ben Pease.

on the continent, making for uninterrupted and relatively uniform—or, to be precise, gradually transitioning—climate conditions across a very wide region. In shaping the mosaic of this place, human inhabitants and nonhuman forces "selected" from many diverse tiles. Moreover, the prairie Midwest had a surface on which many tiles usually could thrive but one that did not hold them very firmly in place. Relative to other "mosaics," then, this is a distinctive one, not only because of its actual shape in any given moment, but because of its instability—the way in which its special ingredients and conditions made possible so many alternative landscapes *in potentia*.[8] The potential and actual flux in the region's ecology was fundamental. Here was a landscape of tension, a volatile and unstable landscape *of* transition, frequently *in* transition.

These transitions—in both senses—were neither fully anthropogenic nor fully nonhuman. A deep understanding of the history of the tallgrass region helps us move past this sharp opposition; the key is to see the entire mosaic as the product of important and specific moments of change or even events. Many of these, of course, had to do with what historian Dan Flores calls the "grand forces": climate, subduction, and the paths of growth and entropy that define the energy flows of the planet in myriad cycles and systems. But when humans first arrived on the scene they became the most important artisans of the mosaic, even as their ability to shape the landscape's configurations was often sharply limited and constrained by such big picture realities. Few people think of this deep history when they see the corn belt's landscape today, but we can try to recover it, for it is crucial to our story. The first step is to get a better and more detailed perspective of the "baseline"—the variable essence of this edgy shoreline landscape in the past.

◆ ◆ ◆

It is no surprise that when French eyewitnesses showed up in the Illinois Valley for the first time in the late 1600s, they were struck by the immensity of the prairies. Prefiguring the discourse of the "sea of grass," explorers registered the scale of these grasslands, as when Jacques Marquette made the rather typical comment that "the prairies extended farther than the eye can see," a description that became practically a refrain for French travelers in the seventeenth and eighteenth centuries. Many observers noted that in addition to their scale, the grasslands were home to distinctive ecological communities, characterized most conspicuously by the charismatic grazers like bison and elk, whose range began right at the top of the Illinois Valley.[9]

But if they often opened their descriptions with reference to the remarkable prairies, European travelers to this region also understood that the true distinctiveness of the Illinois Valley landscape was not merely its huge expanse of grass, but rather its great ecological variation. Spend any time reading the eyewitness descriptions of the Illinois Valley in the late seventeenth and early eighteenth centuries and you will notice a ubiquitous rhetorical construction: "on the one hand . . . and on the other." Traveling through the Illinois in the 1670s, for instance, Jesuit priest Claude Allouez wrote in typical fashion of the great contrasts he saw all over the landscape: "On one side of it is a long stretch of prairie, and on the other a multitude of swamps." The landscape of the Illinois was not homogenous, one observer wrote, but rather was "equally divided into prairies and forests." For his part, Father Claude Dablon agreed that the area around the Jesuits' seventeenth-century Illinois mission settlements was sharply divided into contrasting communities of forest and grassland. As he wrote, "There are prairies three, six, ten, and twenty leagues in length, and three in width, surrounded by forests of the same extent; beyond these, the prairies begin again, so that there is as much of one sort of land as of the other."[10]

Descriptions that highlighted the contrasts in the landscape between grassland, forest, and wetland were almost universal in early accounts from the eastern ecotone region. Jesuit Gabriel Marest wrote that the Illinois Valley was defined by these great contrasting vegetation assemblages, all in close proximity: "Sometimes we have been on prairies stretching farther than the eye could reach, intersected by brooks and rivers, without finding any path which could guide us; sometimes it has been necessary for us to open a passage through dense forests, amid thickets filled with briers and thorns; at other times we have had to go through marshes abounding in mire, in which we sometimes sank waist-deep."[11]

Some years earlier Robert La Salle noted that the eastern prairie region was "a country of boundless prairies, interspersed with forests of high trees." Jacques Marquette added oak savannahs and groves to the list of plant communities he saw alternating with the grasslands, writing from the Peoria village that "one beholds on every side prairies, extending farther than the eye can see, interspersed with groves or with lofty trees." For Canada official Antoine Raudot, the prairies were laced with "patches of woods, with orchards, and with avenues of trees which it seems as if nature took pleasure in making grow in a straight line equally distant one from another."[12]

As Marest's description makes clear, the contrasts in this diverse landscape were revealed especially clearly when an eyewitness narrated the experience of traveling through the region. It is from these frequent descriptions that we get our best view into the nature of the ecotone landscape, which was hardly uniform prairie but rather contained a multitude of different ecozones. To take one more illustrative example, Pierre-Charles de Liette listed five distinct plant communities—including two different forest types—that he saw in the space of just a few miles at the edge of the Illinois' village in the 1690s:

> You see places on the one side that are unwooded prairies requiring only to be turned up by the plow, and on the other side valleys spreading half a league before reaching the hills, which have no trees but walnuts and oaks; and behind these, prairies like those I have just spoken of. Sometimes you travel a league, seeing all this from your boat. Afterwards you find virgin forest on both sides, consisting of tender walnuts, ash, whitewood, Norway maple, cottonwood, a few maples, and grass, taller in places than a man. More than an arpent in the woods you find marshes which in autumn and spring are full of bustards, swans, ducks, cranes, and teals. Ten steps farther on are the hills covered with wood extending about an eighth of a league, from the edge of which are seen prairies of extraordinary extent.[13]

What Liette and others were describing was obviously not a homogeneous grassland, but a patchy, edgy landscape—the forest-prairie mosaic. Here was a mix of diverse forests, grasslands, and wetlands featuring huge populations of birds, grazers, and aquatic animals. Importantly, French eyewitness descriptions are corroborated by the records of the land surveyors who in the early 1800s provided on-the-ground accounts of the landscape cover of the midwestern states, and whose notebook descriptions also foreground the theme of edginess and patchiness across the flatland.[14] It is a much better image for us to carry in our heads than the sea of grass. Yet it is still only a starting point for our understanding of what made this bioregion so distinctive in premodern times. Because while early eyewitnesses could see the great diversity of the landscape, what they could not see was just as important: its dynamic history. More than flatness, more than prairies, more than big sky, the really key feature of the eastern prairie landscape in many ways was flux.

◆ ◆ ◆

All landscapes are variable and dynamic, and the tallgrass prairie mosaic was certainly not unique in this regard. A central truism of ecology since the mid-twentieth century holds that vegetation communities are not static but in constant transformation. This was not always the consensus view, and we should recall a bit of intellectual history to note how recent this understanding of dynamism in ecology really is and to remember that the older ways of viewing still stubbornly persist in certain ideas about the natural world.

Among the founding generation of ecologists, and indeed among others who approached the issue in the twentieth century, it was taken for granted that natural environments had a sort of built-in teleology, a kind of nonrandom tendency governed by what were once understood as laws of ecological succession.[15] Most importantly, many believed there was such a thing as a "climax": a single predestined shape that any given plant community would take over time. In the view of Frederic Clements, one of a founding generation of ecological researchers trained by Charles Bessey at the University of Nebraska, a typical plant community proceeds through fixed successional stages to a monoclimax, and then reaches its destiny as a self-equilibriated "formation." As the fields of ecological management and restoration picked up steam in the midcentury, researchers often used the concept of "climax" as an unproblematic reference point for their work. The climax was also related to the notion of wilderness, a sense that nature "out there"—or separate from humans—had a certain balance and stability. In dedicating the University of Wisconsin's arboretum in 1934, for instance, ecologist Aldo Leopold expressed this understanding of a climax formation as the key concept informing the project. The purpose of the arboretum, he said, was: "The reconstruction of original Wisconsin . . . to serve as a bench mark, a starting point, in the long and laborious job of building a permanent and mutually beneficial relationship between civilized men and a civilized landscape. . . . The first step is to reconstruct a sample of what we had to start with."[16]

To Leopold and many others this notion of "what we had to start with" was solid and stable. Just like any organism—or perhaps "superorganism"—any given environment reaches maturity and flourishes as the climax version of itself if left alone and allowed to thrive. Within the emerging discipline of historical ecology, setting the "baseline" for landscape change was a simple matter of identifying what climax configuration was "natural" for a given landform, owing to climate conditions, geology, and other factors.

In the latter half of the twentieth century, however, ecologists fundamentally changed this assumption. First, they questioned the idea that plant communities behaved as functioning organisms, and instead emphasized how individual plant species associated together "mechanistically"—and even somewhat randomly—in a given habitat. In this context, many abandoned the idea that plant communities—now understood less as a coherent or distinctive "association" and more in terms of H. G. Tansley's notion of "ecosystem"—function with any kind of inherent, regular tendency toward climax.[17] Other changes followed. Moving away from the notion of the climax, researchers interested in the dynamics of ecosystems over time increasingly looked not for a static reference point, but rather a "historical range of variation" that defined a much broader and more dynamic "baseline" for past ecosystems. They emphasized not a linear trajectory toward climax, but instead a very much nonlinear set of processes—including a dynamic of disequilibrium—that over time produces varieties of landscape configurations. As landscape ecology, ecosystem management, and historical ecology continued to develop as coherent disciplines, their practitioners recognized that identifying "what we had to start with" in any given landscape is actually a historical question dependent on the complicated and subjective question of *when*, in fact, we started.[18] Moving away from a static sense of "primordial" nature, historical ecologists defined reference points such as "potential natural vegetation"—or the kinds of vegetation communities that would predominate in the absence of human influence—even as they recognized that few landscapes ever met this "undisturbed" condition in the Holocene. Indeed, the increasing recognition of human influence in many supposedly "wild" landscapes opened up new fascinating angles into historical ecology research, but it also substantially complicated the previously simple notion of the "natural" baseline reference.[19]

All this history informs an important point about the past landscape and natural environment of the tallgrass prairie region. Because if all landscapes are in transition and should be thought of in terms of a range of variability, it is nonetheless true that the tallgrass prairie bioregion was especially dynamic. As ecologists have long recognized, the flux and instability of the tallgrass region were simply more intense than what many other regions experienced in the Holocene. Indeed, the variability of the region was so intense that when

August Wilhelm Küchler defined the biozone of the eastern prairie, he built variability right into the essence of the place.[20] Rather than a consistent zone, Küchler defined the region as a transitioning mosaic. Just like the French eyewitnesses (although with different words), his definition noted the *variety* of the ecozone. Importantly, Küchler's definition also built in a notion of flux and change, or the notion that any given expression or configuration of the mosaic is just one possibility within the range of "potential natural vegetation." To borrow Leopold's expression, the essence of *what we had to start with* in the eastern prairie region was in fact instability. But why?

The shape and functioning of any given ecological community—including the tallgrass ecotone—is at bottom the result of the interaction of some basic metabolic principles with the specific features of the local landform. To get us thinking on the right scale, it is helpful to remember first of all that the earth is an energy system, powered to a small extent by the heat inside its core but mostly by the sun. The sun constantly bombards the planet with energy, and a small portion of this energy (for most of it dissipates unused) powers virtually all of the motion and dynamism on this planet directly through radiation and convection and indirectly through chemical conversions and other energy transfers. In the big picture, ecology is simply the metabolism of this energy.[21] A tiny fraction of the sun's energy flow (less than 1 percent) is absorbed by pigments in plants energizing photosynthesis; a larger part warms the bodies of plants and animals and their shelters. Meanwhile, radiation heats the oceans and rocks and soils and drives critical biospheric processes: it powers the water cycle, generates winds, and maintains temperature ranges suitable to plant and animal metabolism and organic decomposition. Related to each of these processes, this solar radiation also indirectly powers climate, which then produces the weathering of soils and the mobilization of minerals needed for primary production in the overall metabolism.[22]

In the biggest sense, the prairie peninsula took shape in history because specific geological features shaped specific climatic patterns in which distinctive organisms, chemical processes, and thus energy pathways developed. Given particular developments in landforms and changes in the ways in which certain plant producers synthesized chemical energy in the far distant past, different living assemblages began to compete for dominance in this region over long periods of time. It is this notion of competition which is central. Going against his own concept of climax ecology, Clements summed up the essence of the eastern prairie zone as a place where broad cycles of climate variability and ecological change produced a diverse collection of grassland and forest species but "no clear victor" among plant communities contending at the edges.[23] For natural historian John Madson, the eastern prairie region in the Holocene was even more explicitly a "battleground of floristic groups," where the "meeting ground of North American weathers" enabled constant "advances and retreats" by different ecological assemblages.[24] Several key deep-time events are responsible for these advances and retreats. The first part of the story—the first *event* in how it came to be this way—has to do with mountains. As Fernand Braudel observed in his famous study of the Mediterranean—which is no less true for the midcontinent of North America—"mountains come first."[25]

◆ ◆ ◆

Beginning some sixty-six million years ago, the so-called Laramide Revolution pushed a massive expanse of low-lying western plains up some three miles above sea level, a monumental height from which erosive factors have steadily reduced them ever since. As soon as they rose up, the resulting Rocky Mountain range blocked moist Pacific air masses—westerlies that previously carried precipitation-forming clouds from the ocean onto the land—and "wrung out" the moisture, leaving on the eastern side a huge dry region that stretched for hundreds of miles. The "rain shadow" effect was strongest closest to the mountains, while other sources of precipitation (from the north and south) lessened its severity further to the east. This created a vast precipitation gradient across a huge expanse of the midcontinent, with annual rainfall averages ranging from eleven inches at the immediate threshold of the mountains to forty-seven inches on the rain shadow's easternmost edge. The Laramide Revolution made one of the most important "breaks in continental climate."[26] It fundamentally shaped the rise of the ecotone. Of course, on one level the reasons for this are obvious and intuitive: differential rainfall supports different vegetation and can tip the balance of Clements's "competition" toward one or another vegetation group. It makes a gradient, not just of species type but of overall primary productivity, so that overall photosynthetic activity becomes more intense—indeed, becomes quite rambunctious—toward the eastern edge of the zone. The "true prairie"—the tallgrass prairie—eventually took shape at this easternmost edge of the rain shadow, where the annual precipitation was over forty inches and where rates of evaporation and precipitation were nearly equal.[27]

The Western Cordillera shaped precipitation in the land to its east, not just from year to year but also on a seasonal basis. As a result, the region is distinctly deficient in wintertime precipitation. Big storms saturate parts of the easternmost part of the rain shadow in the spring and summer, but not in the winter. Meanwhile, the interaction of the mountains with cyclonic currents—the earth's wobble, sunspot cycles, El Niño–Southern Oscillation (ENSO) events, and other more long-term "phases"—also produced dynamics and variability on a broader temporal scale of decades and eras. For this reason the region of the rain shadow has been especially sensitive to climate events, owing to the ways that the mountain range affect changing air currents, or the "whipsawing of climate" throughout the region. The most frequent and important climate effect is drought; in much of the prairie peninsula a drought lasting more than three years has occurred every 80–120 years or so. In other words, the mountains produce a somewhat steady east-to-west precipitation gradient, but they also produce short-term volatility. As is true in many ecological contexts, extremes of climate in the rain shadow have often been more important than climate averages in determining the ecological competition that took shape there, since extremes provide the limiting conditions for certain species to thrive.[28] This brings us to the next important event in our story, an event that is much more squarely biological.

Well after the uplift of the Rocky Mountains, a microscopic change, equally momentous, took place in the physiology of certain plants: the evolution of the so-called C4 carbon pathway. Grasses, or members of the family *poaceae*, which had begun to evolve in tropical conditions around the same time as the Laramide Revolution, evolved with many distinctive characteristics, including, most notably, a distinctive "spikelet" flowering structure and a low growth point as an adaptation to grazing animals. The C4 carbon pathway that accompanied these changes refers to the four carbon acids that these plants are able to concentrate or "fix" as biomass. The C4 plants evolved high-water-use efficiency, stomatal sensitivity to water loss, and high-photosynthetic rates. They were more efficient and productive in periods of high-moisture stress, and did much better than C3 plants under conditions of high irradiance and temperature. The development of the C4 pathway was a crucial event in primary biological production on the earth, with C4 plants eventually accounting for some 20–30 percent of the carbon fixation on the planet and over time leading to the rise of the world's great grasslands and savannas. Eventually this transformation would be critical to agriculture, since 30 percent of global grain production is of C4 plants.[29]

To understand why the C4 evolution was so important to the ecotone, however, we should think of plant assemblages not in terms of static zones, but rather of long-term "migrations." A truism of the discipline of ecology is summed up by pioneering botanist Henry Allen Gleason, in his 1922 opus, "The Vegetational History of the Midwest": "The history of the vegetation of the Middle West, as of every other portion of our continent, is a history of repeated migrations of diverse floristic elements, arriving in the region from various directions, persisting there for various lengths of time, and finally retreating under the pressure of environmental changes which made their position no longer tenable."[30] Plant migrations are a matter of competition. Plants are opportunistic, and they crowd each other out in processes of constant invasion and retreat. Climate conditions, as Clements acknowledges, are thus only part of the story and will "decide the fate of the competitors *in accordance with lifeform*" (emphasis added).[31] For Gleason, who differed from Clements on this point, plant communities do not act together as a superorganism; rather, they compete individualistically with one another to shape and change resulting vegetational assemblages.

Here, then, is why the rise of C4 grasses represent such a "revolution" in the shaping of the ecotone. The drought-prone conditions created by the shadow of the Rocky Mountains over millions of years eventually created space for the rise of one of the world's great grasslands in North America, where C4 plants dominated. Other things being equal, C4 grasses could outcompete woody species, trees, forbs, herbs, and other grasses that used the older C3 photosynthetic pathway, particularly in seasons of drought. In tallgrass prairie, C4 grasses eventually would come to comprise over 80 percent of aboveground net productivity (and about the same percentage belowground), despite representing a minority of the overall five hundred species in the assemblage.[32] To be sure, the evolution of C4 grasses by itself was not determinative of the prairie peninsula. To the contrary, paleoclimate research makes it clear that grasses were only sometimes present in the midcontinent in the millions of years since they first appeared and developed their competitive advantage. But their evolution became an important potential ingredient—one of the many key circumstances that would define the region's flux and ecological "competition" in more recent times—as the grasses' particular advantages intersected with other contingencies of deep time.[33]

A third and related "revolution" in the shaping of the prairie peninsula has to do with the creation of the most physically conspicuous of all the prairie peninsula region's characteristic features: flatness. At the beginning of the Pleistocene epoch, giant ice sheets from the arctic began a series of advances south into North America, rearranging the geological features of the northern half of the continent. In the future Midwest, glaciers advanced and retreated several times, covering almost the entire region and leaving behind a distinctively flat surface covered with glacial "drift" (the finely ground-up rock and clay that settled in the wake of glacial retreat). The glaciers in the Midwest interacted with a surface that was already relatively flat, having been covered by an inland sea for much of Paleozoic times (beginning five hundred million years before present [mybp]). In the eastern prairie peninsula, glaciation "finished" the land like an iron would, filling in valleys and scouring off promontories. The most recent glaciations of two hundred to eighteen thousand years ago left behind distinctive glacial moraines and kettles in many parts of the eastern prairie peninsula. But mostly they left behind a super-flat surface, clayey soils, and slow-moving drainage into watersheds like the reconfigured Illinois River Valley.[34]

The upshot of the glaciation in the Midwest was dramatic. In many places in Illinois, for example, the land relief in the state is less than 3 meters per kilometer. The total relief in the state is about 290m, with typical elevations ranging between 180m and 240m above sea level. Flatness is, of course, the region's most distinctive characteristic, and much maligned. But few appreciate how the flattening of the midwestern landscape fundamentally shaped ecology in ways both active and passive. On the active side, the glaciation produced important dynamics for hydrology in the Midwest. To put it bluntly, water simply doesn't move across flat lands, and the combination of this fact together with the "tight" clay soils of glacial drift led to many parts of the glacier-flattened midwestern landscape being swampy and wet for much of the year. Despite our frequent assumption that the tallgrass prairie was a mostly dry landscape, in fact the only xeric—or dry—prairies in the eastern prairie peninsula in the 1800s existed on well-drained soils or on areas of steep relief, which the glaciers had made rare.[35] In many areas of the prairie from Illinois north to Minnesota, for instance, 40–60 percent of the land cover was wetland for much of the year, particularly the spring. While agricultural history rightly gives John Deere's sod-busting steel plow pride of place in the narrative of the region's agricultural development in the nineteenth century, poorly drained soils were just as big an obstacle to would-be farmers as the dense sod and tough root systems of the prairie landscape. Indeed, it took thousands of miles of subsurface drainage tiling to eliminate the wetlands of the Midwest and convert the land to row crop agriculture, beginning in the late-nineteenth century. In the Illinois Valley, a rambunctious and changing floodplain characterized by backwater lakes, sloughs, and flooding made a "watery world" that greatly affected ecological development. The Illinois River itself is conspicuously flat, draining at a rate of around 20cm per km in the upper reaches of the valley and as little as 2cm per km in its lower stretches.[36] While perceived as a liability to would-be settler colonists in the modern period, the region's hydrology added conspicuous biodiversity to the landscape and, even more important, *productivity.* Of all the world's biozones, wetlands are among the most efficient—second only to tropical forests—at concentrating sunlight into phytomass. By actively applying the feature of flatness to the midwestern landscape, glaciers contributed a central ingredient to its ecological potential.[37]

While glaciers applied certain characteristics, however, equally important were the landscape qualities that the glaciers might be thought to have withheld, or passively removed. As ecologist Paul G. Risser writes, ecotones and ecological transition zones occur under two types of conditions. The first is a sharp break, such as a sudden change or steep gradient in climate or other characteristics of the land: soil type, hydrology, sunlight availability, or especially topography. A good example of this first type of resulting ecotone is the area between a plain and a suddenly rising mountain range. Such a landscape transition creates a sharp discontinuity in ecological dynamics within a relatively compressed space and shapes a distinct and unavoidable change in floristic communities on the land. The other kind of ecotone has no sharp break and instead is characterized by a blurry "threshold." In this kind of situation, vegetation communities and fauna exhibit much more gradual and patchy interaction across the zone, based on (in Risser's terms) "nonlinear responses to gradual gradients in the physical environment that cause large changes in ecosystem dynamics and the distributions of dominant species."[38]

By removing all the topographical landscape features that might have created steep gradients, glaciers passively shaped the broad transition in the future corn belt, setting up these "nonlinear" dynamics. In the context of the aforementioned processes of vegetation migration and competition, glaciers and the landscape's resulting flatness ensured that, with the obvious and important exception of water-saturated areas, there were few clear and unavoidable physical limits on many species' ranges across the dynamic zone of the rain shadow, with its gradual rainfall gradient and intense climate variability. Over 60 percent of the modern state of Illinois has a relief of between 2 and 4 percent. Additionally, well over half of the state's land area is covered by soil without any limiting morphological feature such as claypan or shallow bedrock. In this context, vegetation competition was never a matter of a sharp boundary line or a fixed limitation of any kind. Even if C4 grasses could have grown more efficiently in drought, it seems clear that this was not always determinative for the ecology of the region, at least not always in a stark way, since drought does not distribute in a regular pattern across the land. And, if drought was not always a decisive factor, annual precipitation was even less determinative, particularly at the eastern edge of the rain shadow, where rainfall averages were easily sufficient to support many species of deciduous trees, C3 forbs, and C4 grasses. As Carl Sauer provocatively notes, paleoclimate evidence shows that climate itself was not enough to explain the rise of grasslands, or their location, on the flatland.[39] In Clements's terms, changing cycles produced "periods of varying duration" in which certain plant communities could "hold the ground won by the favor of the changing cycle," but only for a time.[40] With the conspicuous exception of its effect on river hydrology, then, the glaciation of the midwestern landscape acted passively to withhold determining landscape features like topography and soil type that in other regions created sharp ecological boundaries and vegetation zones.[41]

In the end, the Laramide orogeny, the evolution of C4 grasses, and the Pleistocene ice ages were the crucial deep-time events underlying the special conditions at the prairie-forest border. Charting the rise of these special and important landscape features gives us an initial sense of why the landscape was so unstable and dynamic, and helps us begin to understand its potential. Closer to the present, on a still grand but more human time line, all of the crucial deep-time environmental conditions interacted with certain dynamic processes on the ground to shape the region.

◆ ◆ ◆

After the glaciers, the environment of the prairies slowly settled into a warming climate that is recognizable. Pollen records from Illinois locations like Chatsworth Bog and Volo Bog show that after the ice receded, tundra vegetation was the first to be established, then spruce forests around ten thousand years ago. Subsequently came mesic forests, which had become dominant in the region after the decline of spruce changed the composition of the forest into oak-hickory, the characteristic forests of the uplands Midwest at present. Then, beginning around eighty-three hundred years ago, immigrant grasses began to invade, moving in from refuges in the south and southeast, where they had waited out the Ice Ages. Evidence suggests that these grasses were coming together in the Midwest for the first time, not as a "reconstitution" of a previous ecosystem but as a novel vegetation assemblage. Signs of this lie in the fact that the tallgrass prairie assemblage was mostly comprised of species that had evolved in other ecological configurations, with few "endemic" species unto itself. If this suggests the contingency of the tallgrass's creation in this relatively recent period of time, however, the prairie was nevertheless soon a robust biozone, and growing. Between eight and six thousand years ago the prairie gradually pushed east, as an extensive hot and dry period—the so-called Altithermal or Hypsithermal—hit the Midwest. This gradually evolving prairie spread east almost as far as modern-day Indianapolis, developing a distinctively "patchy" southern border.[42]

This eastern push of the prairie was a maximum and may represent the furthest ever extent of the newly evolved and unstable ecosystem.[43] A few more periodic and episodic climate events—for instance, most recently the warm and dry period of one thousand years ago, the so-called Medieval Warm Period—may have been instrumental in supporting the prairie's easternmost push and maintaining it. But for the most part, climate conditions were acting to reduce prairie extent—and to favor forest succession—for much of the period in the last five thousand years.[44] Extending the shoreline metaphor, we might think of this as analogous to a tide, or a maximal inrush of the grassland "shore," which now was preparing to run back out. Another way of thinking about it is in terms of Madson's "battleground of floristic groups": the grasslands in 5000 BCE had pushed as far east as they ever would, and now the forest species were pushing back. The potential energy of the system—at least as far as climate was concerned—seemed to be trending toward forest.[45]

In this context, however, another process helped to shape the region's dynamics: fire.[46] In certain ways fire is the most ancient actor in this discussion, even older than the Laramide orogeny. Fires have been smoldering on the planet for four hundred million years. Fire has been a huge force in the metabolism of the planet, a "biochemical flywheel" hastening the breakdown of phytomass and redirecting the carbon cycle through specific energy pathways.[47] But even though fires are ancient, in the eastern prairie region they were also somewhat new. Pollen and charcoal records from after the glaciers reveal how a pattern of regular fires was established in the eastern prairie region, coincident with the eastern push of the prairies in droughty

Fire on Nachusa Grasslands, Lee County, Illinois, 2013. Photo courtesy Charles Larry.

conditions some eight thousand years ago.[48] Both the cause and the effect of the extension of the grasses, the increase in the number of fires shown in charcoal records is clearly a big part of the story of the eastern prairies.[49]

Without trying to resolve whether particular fires of the past were anthropogenic or wild, it is enough here simply to note a few principles of fire history. A fire regime, whether anthropogenic or lightning-driven, needs cured fuel. In many if not most temperate landscapes, the limiting factor for fire is not fuel but rather ignition.[50] This has certainly been true in the eastern prairie region, where primary productivity in the eastern edge of the rain shadow is high and fuel accumulates quickly. Not lacking for fuel, what fire needed in order to establish a regular regime was a felicitous wet-dry cycle. It is not the averages—the forty, thirty, twenty, or even ten inches of annual rainfall—that determined whether any part of the landscape was going to support fire and in what form. Rather, it was the "peaks and pulses," the rhythms of wetting and drying.[51] Dense canopied forests are able to retain their moisture and thus are difficult for fire to "crack open." But in a shift that probably represents both cause and effect, the push of C4 grasses into the eastern prairie region during the Holocene dramatically increased the frequency of fire in this landscape, probably by providing a wholly new kind of fuel that was much more susceptible to ignition. Importantly, the C4 grasses established a particular cadence of moisture exchange, greedily absorbing and channeling water into rambunctious spring growth before curing and drying in the summer and fall. Indeed, certain ecologists have posited that the most important eastern advance of the C_4 grasses in the recent past (eight thousand years ago) was owing not so much to decreased average annual rainfall, but rather to a new "patchy" climate that encouraged fire and to which the C_4 growth pattern was especially well adapted. This monsoon climate may have been accompanied by frequent lightning strikes at the end of the dry season; it was certainly accompanied by human occupation. Whatever the source of ignition, however, fire was a primary driver in the conversion of forest to grasslands and then the maintenance of those grasslands.[52]

Incidentally, fire was especially powerful in previously glaciated landscapes of the prairie peninsula, because fire has an important interaction with flatness. If the flat landscape creates no stark topography to separate vegetation communities by growing conditions, it also creates no fire breaks to stop or dampen fires once started.[53] It is easy to intuit the important mechanism here. Owing to convection and the tendency for flames to "run up" the windward slopes of hills before dying at the top, stark elevation changes can make a big difference in the progress and effects of big fires. In places like modern-day Illinois, where the elevation changes are subtle, historic forests survived most frequently on land with average slope ranges greater than 4 percent. But in most of the flat landscape, that great portion of the land with elevation change somewhere between 2 and 4 percent, fires could run on without interruption.[54] Flatness worked to enhance the effects both of applying and of withholding fire from the landscape, even as it also directly determined vegetation composition through regulation of soil moisture.[55]

To be sure, the question of just how Holocene fires were ignited is not yet our focus. For some ecologists it is enough to posit that lightning was frequent in the Midwest. Others point out that most lightning in the eastern prairie region comes with thunderstorms and, by definition, during the rainy season when vegetation is too wet to carry fire for long. It is a thorny problem that should be approached specifically, not generally. But regardless of how fires were ignited, we know they happened, and when they did they had a profound ecological effect. Fires burned back woody species at high rates; one recent fire in central Illinois killed 34 percent of the relatively fire-hardy oak and hickory trees within a year.[56] Furthermore, in the aftermath of fires, tree seedlings are easily outcompeted by grasses. And, in addition to shading out small tree seedlings and preventing forest regrowth, grasses also produce flammable, finely divided fuels that encourage the spread of future fires to which trees are again vulnerable.[57] Of course, on the other hand, in the absence of fire, forests can reclaim and maintain a landscape relatively impervious to ignition. Fire is a key variable, whether present or absent.

The final dynamic process in the Holocene history of the prairie peninsula involved animals. If fire altered vegetation and thus helped direct the production of biomass through specific photosynthetic pathways, grazers had a similar effect in the eastern prairie region when they were present. Unlike fire, grazing is not likely to move a forest boundary—grazing animals do not have the capacity to kill large trees. But grazers most definitely could maintain grasslands once established, through several main functions. First, by removing large quantities of biomass from the landscape, they could prevent the overgrowth of certain dominant C4 species and thus enhance the biodiversity and stability of the prairie, effectively holding the line against forest succession. Relatedly, by trampling the vegetation, grazers could help activate and stimulate growth in grasses and forbs and kill woody species. This grazing also had the effect of reducing the biomass in the grasslands, lessening the severity of fire in certain grazed patches and promoting regeneration of fast-growing grass species. Ranging across the landscape, grazers' own metabolic processes also had the ability to redistribute nutrients and add fertility to the landscape, adding to variety and productivity.[58]

As in the case of fire's interaction with flatness, grazers' influence worked intersectionally with fire. More specifically, the effect of grazers and fire on tallgrass prairie ecosystems was mutually enhancing and perhaps even mutually regulating. Given the high productivity of grassland ecosystems, particularly in zones of high moisture, a good portion of the season C4 grasses often produced too much cellulose to be palatable for grazers like bison and elk. But fire could direct and shape the biomass of the prairie seasonally, causing grasses to send up fresh leaves that attracted grazing animals, whose grazing in turn affected fire ecology. In its application and in its removal, in other words, grazing had the power to fundamentally direct the ecology of the prairie-forest edge, both by holding back succession and through its interaction with fire.[59]

Like fire, the presence or absence of grazers is less a principle and more a cycle, or even an event, and we must approach it specifically as a matter of history. For now, however, it is enough to note that big grazers like giant bison are present in the fossil record of the eastern prairies at various times and absent at others. Indeed, positioned at the edge of the rain shadow, the eastern prairie region was often the eastern edge of what historian Dan Flores calls the "American Serengeti": both the old Pleistocene version, with its massive and strange megafauna, and the more recent Holocene version, featuring bison, elk, wolves, and other fauna. In certain parts of the tallgrass ecotone, bison and elk may have become "keystone species" at various times, although it might be better to say generally that at various moments animals and grasses together established a kind of "grazing regime" akin to the "fire regime" already discussed. Whether present or absent, they were part of the flux of the landscape on a human scale.

Given the importance of these dynamic elements—short-term climate events, fire, and grazers—in shaping ecotone boundaries in the prairie Midwest, it goes without saying that people could control the latter two but not the former, when they arrived in the eastern prairie landscape some twelve thousand years ago. Arriving in this special place, they likely quickly learned that the dynamic elements of the landscape created both possibilities and limits for human history. Over time people could use fire to promote game or to hunt by driving animals to a kill zone. By encouraging, preserving, or eliminating herds of grazers and other game, people could use animal populations to shape the fire patterns of their regions. Relatedly, people could wrangle fire to shape landscapes on both big and small scales, by applying and withholding it to manage game, create productive landscapes, or perhaps set boundaries. Climate affected their ability to do all of these things.

Through it all, short- and long-term patterns contributed to the great flux in the landscape, driving dynamic processes of diversity and change. With its special characteristics, the landscape remained contingent, and though many of the patterns discussed here were long term and slow moving, it is also true that these realities could sometimes produce relatively quick changes. Recent research on ecotones—including on the forest-prairie edge in North America—has emphasized the surprisingly rapid responses of ecological processes to changing climate conditions. For instance, in addition to slow and gradual migration, forest edges can also move quite starkly and quickly in response to specific disturbances in climate, fire regime, and grazing patterns, and on human-scale time lines.[60] Recent research shows average tree ranges presently moving to the west over midwestern North America at some 30km per decade, a transformation likely driven by climate change and the absence of fire.[61] In part because of its late evolution in the Holocene and in part because of geologic conditions, the new plant assemblage of this shoreline mosaic was by definition unstable and subject to fragmentation and invasion.[62] As Clements says, it was a battleground, a constant competition among diverse floristic groups, with "no clear victor." Not homogeneity, but rather flux, was its true essence.

◆ ◆ ◆

Beginning twelve thousand years ago, the most important landscape change in the prairie ecotone started with the arrival of people. In settling the region, people engaged with this landscape of flux, with its peculiar energy streams, and began a long history of interaction. Early hunters made their lives with giant megafauna, as witnessed by large caches of Clovis points found in the Midwest.[63] Groups likely migrating from the south made the first sedentary societies in the Illinois River Valley during the Archaic period, taking advantage of the extensive wetlands shaped by river aggradation in the Hypsithermal and possibly the climate-forced "degradation" of the prairie uplands environment.[64] Early horticulturalists invented a distinctive complex of domesticated plants as they practiced an increasingly intensive seasonal round, concentrating their subsistence on a broad spectrum of resources in the Illinois River Valley, whether by opportunity or by necessity.[65] Later, fully agricultural societies known as Hopewell rose and fell in the midst of a climate shift (around 200 BCE—200 CE), which may have enabled them to concentrate more narrowly on the special crops of the midwestern agricultural complex—plants like goosefoot, sumpweed, and little barley.[66] Much closer to the present, the Midwest's abundance supported the rise and fall of a great maize-based agricultural civilization—the Mississippian climax of Cahokia—in a moment of climate change.[67]

The relationship between people and the special regional environment over this long sweep of history is too complex a process to be simplified or generalized. But significant shifts in climate and environment—major episodes of drought, of river aggradation, and the like—frequently shaped human history in this sensitive region. Simple renderings of the prairie's natural history consider it as a monocultural sea of grass, a static and uniform "climax." Meanwhile, often overly focused on the question of whether the prairie region was a fully anthropogenic landscape, many ecologists of the prairie region have ignored a more important story, the special background characteristics of the landscape that made it so prone to change. This region was distinctive: newly evolved, climate sensitive, exceptionally flat, and at the meeting ground of two of North America's biomes. Its essence was tremendous diversity, as well as instability and flux.

Deep history can help us understand why. More like a shoreline than a sea of grass, this ecotone moved and extended and receded, changing together with its human inhabitants. Far from a static stage on which dramatic events would play out, then, this ecotone was part of those changes and part of the story. During what we might consider North America's medieval period, specific forces out of the deep past were about to shape a specific historical contingency: the rise of an unsung nonhuman actor of midwestern history.

2

SPECIES SHIFT

NO ONE WOULD ACCUSE THE CITY OF SPRINGFIELD, ILLINOIS, OF ignoring the past. Arriving in Springfield on any major road, travelers encounter numerous signs directing them to historic sites, especially ones related to Abraham Lincoln—his tomb, the Lincoln Home National Historic Site, the impressive Abraham Lincoln Presidential Library and Museum. At many of these attractions, dedicated public historians and volunteers stage robust educational programming, informative tours, and enthusiastic reenactments about important events of local history and memory.

There is one true locus of historical significance in Springfield, however, about which there is less roadside signage, and surely altogether less consciousness let alone fanfare. In a large but bland building tucked away next to a railroad track and behind a tall fence near the center of town is the Illinois State Museum's Research and Collections Center. Featuring offices for scientists in disciplines like paleontology and geology, much of the building's interior is no more remarkable than a typical American public-sector office space, albeit with a jarring mix of state-of-the-art computers and lab equipment resting on outdated workplace furniture. But walk through double doors at the back of these offices into a large, dimly lit room in the deep interior of the squat building and you enter into what is the most extraordinary assemblage of history in the whole midwestern region, indeed one of the most significant specimen collections east of the Mississippi River. Featuring 14.5 million objects, the massive collection room contains everything from microscopic pollen grains to gigantic mastodon fossils to carefully preserved plant specimens and bits of pottery. Scattered around are traces of the past spanning literally millions of years. Walled off from Springfield's other sites of historical commemoration, by both literal walls and disciplinary traditions that separate the human and nonhuman parts of the past into discrete and often unconnected stories, these materials call out to be reintegrated into our sense of history in this place.

One particular candidate for such reintegration sits on an oversized shelf in the main collections room in large gray boxes resting on foam pads. Here the curators of this incredible collection preserve several specimens of an important actor whose historical significance is not yet recognized on the same level as Abraham Lincoln's: the bison. From the enormous skull of the ancient *Bison latifrons* (an ancestor of the modern species from the mid-Pleistocene) to the horns of the *Bison antiquus* (another ancestor that went extinct some ten thousand years ago) to the impressive leg bones of the modern *Bison bison* (which lived and died in much more recent, albeit still remote, times), the old bones, skulls, teeth, and other semifossilized body parts are among the most charismatic "objects" in the collections room, as stunning as the lifelike taxidermy in the adjacent section of the storage facility. Of course, these are also *bones*, and they have none of the benefit of historical reenactors enthusiastically bringing them to life. Given this limitation, and considering them in the context of this natural history collection which everywhere contains reminders of the long, slow processes of evolutionary change, it may be hard to think of these animals as being part of historical events at all, let alone on the scale we usually imagine. Yet they were indeed a crucial part of an important moment of change, a rupture with distinctive consequences for human history.

In the late twelfth century a major climate event, part of the continental climate disruption known as the Pacific Drought, struck the Midwest.[1] Simultaneously, and no doubt with some degree of connection, the massive chiefdom of Cahokia, with its thousands of people organized into city-sized settlements around great ceremonial plazas, began to break apart. The ensuing generations witnessed a set of human migrations, as diverse people moved both into and out of the midwestern region. Slow and probably subtle in many of their effects, these climate changes and shifts in human geography nonetheless created the context for one relatively abrupt and less-noted change in the faunal life of the eastern prairie region: the appearance of large herds of bison grazing in the eastern grasslands. For the first time in the Holocene, herds of animals numbering in the hundreds and thousands proliferated especially in the prairie uplands, probably by the 1500s. Before long, people here began adopting a bison-centric lifeway, making lives together with these huge animals and the energy streams they embodied. Here was a true and relatively rapid revolution in the human ecology of this part of North America, one that reflected the special flux of the tallgrass prairie region and its responsiveness to changing conditions.

Building on what we already know of the instability of this edgy regional landscape, this chapter centers the rise of these grazing bison herds and their human counterparts as a historical event. Drawing on archaeology and important new research on the biogeography of bison in the Midwest, the chapter considers the bison boom in the prairies in this era as an example of a *species shift*. Coined by William Cronon, Jay Gitlin, and George Miles, the term usually refers to now-familiar dynamics of many colonial histories, when introduced and invasive plants and animals moved into new ecosystems and proliferated.[2] In many cases the absence of natural predators for these invaders made for dramatic ecological change; for instance, the runaway growth in populations of viruses, weeds, and certain domesticated and wild animals in colonial contexts. Early American historians have identified many such moments as the centerpieces of a consequential "Columbian Exchange," as well as the cornerstone of the "changes in the land" that made nonhuman animals, plants, and other biological phenomena among the most consequential of all historical actors in the years after 1492. The species shift in many places was the most significant kind of rupture of all.[3]

If we're used to thinking about these kinds of biological events as transformative moments in the context of *colonial* history, however, we must acknowledge species shifts could and did happen in other times and places. The rise of bison herds in the eastern prairies, while nowhere as dramatic in its consequences as the introduction of pigs into the contact-era Caribbean or smallpox into New England, was nonetheless also a significant episode, and for some of the same reasons. Proliferating in an open niche in these eastern grasslands, the bison created a sudden opportunity for people on the move, enabling a human-animal partnership that created new power in a moment of invasion. To understand this we need to reconsider those important historical agents whose semifossilized bones lie in storage in Springfield, silent and motionless but still with a story to tell. The story begins with a decidedly non-bison-centric people, the Mississippians of Cahokia.

◆ ◆ ◆

Like the great mounds that still rise massively over the landscape in modern-day Collinsville, Illinois, the chiefdom of Cahokia looms large in any Indigenous history of the early American Midwest. Starting with a "big bang" around the year 1000, the city of Cahokia rose as an unprecedented social and ecological innovation, a massive corn-based agricultural economy marshalling the labor of diverse peoples in an urban center of at least twenty-thousand residents and thousands more in associated hinterland sites.[4] Intensively hierarchical, the society featured a priestly class of chiefs who ruled over subordinates and enforced hereditary succession, legitimating inequalities through a ceremonial complex that gave elites special access to power from the Above World and the Below World.[5] Surrounded by massive palisades (reflecting the ever-present conflict and strife) and featuring massive mortuary mounds (which embodied the great inequalities of the society), Cahokia was simultaneously a high-water mark of Indigenous cultural achievement and a society rife with tension and division.[6]

While often imagined as a simple story of human power and technological triumph, Cahokia's rise was closely entangled with an important nonhuman entity: *maize*. First cultivated in Mexico, the corn that midwestern farmers adopted around 750 CE was much more demanding than the agricultural crops farmers had previously cultivated in this region, and its requirements shaped the labor regime and much else about Mississippian civilization.[7] A high-productivity but also a high-variability crop, maize interacted with important climate changes in the middle of the continent, both of which enabled and shaped Cahokia's rise.[8] As archaeologists and climate scientists have recently shown, the turn of the second millennium witnessed the onset of a warm period, which enabled the initial turn toward large-scale maize agriculture and powered Cahokia's rise and expansion. But an accompanying wet climate in the so-called Lohman Phase, the moment when Cahokians reached their highest population densities in the concentrated bottomland villages, probably produced poor harvests in the oversaturated and thus anoxic bottomland soils. In response, Cahokia incorporated upland farmers, whose soils were less affected by the wet climate anomaly, into its orbit. This ability to marshal energy from distinctively different parts of the ecotone environment, including both the bottomlands and a crucially important uplands agricultural zone, produced Cahokia's vast power from both a caloric and a symbolic standpoint.[9]

It did not last. The same flux of the region's environment that for a while had interacted positively with Cahokia's corn economy and social expansion also helped bring about its end. After the wet period that characterized the rise of Cahokia in the Lohman Phase, the period from 1100 to 1245 was characterized by "episodic but persistent" drought. In particular, climate proxies suggest that the region of west central Illinois experienced severe drought 140 out of 145 years following 1100 CE.[10] Some or most of the sloughs and marshes in the bottomland region around Cahokia's center may have desiccated, a possibility supported by archaeological deposits that contain fewer fish at the start of this period. Paradoxically, the arid conditions may have created more intense flood events in the bottomlands, which may have been exacerbated by deforestation and resulting erosion along the riverbanks and led to damaging consequences for bottomland dwellers and their agriculture. But the far more important consequence of the onset of drought was felt in the hinterlands. The high water table of 1050 had been a central factor helping upland farmers thrive. Now, with the situation reversed, farmers unable to grow corn abandoned the uplands, perhaps inspired by an especially intense fifteen-year drought beginning around 1100. As this drought continued, the upland Richland agricultural center, which for two generations had served as the "glue" holding Cahokia together in material and symbolic ways, emptied out.[11]

In a US story that too often ignores precontact Indigenous history or oversimplifies precontact Indians with caricatures of "failed farmers" and "lost opportunities," Cahokia's complex story helps us think differently about Indigenous societies. Cahokia was dynamic, powerful, and prosperous, and it often stands as a literal textbook example of Indigenous "sophistication" in the early pages of colonial history surveys.[12] Yet, for all the ways in which the case of Cahokia helps scholars tell a richer history about Indigenous Americans before contact, we still have problems interpreting it. Many popular treatments emphasize Cahokia's supposed "collapse," framing the significance of Cahokia's history in terms of alleged "mistakes" that not only ended the society's achievements but also negated them.[13] Moreover, many accounts take the fact of a lowered regional population in the years after Cahokia's dissolution to be an exclamation point on the chiefdom's trajectory, a "sense of finality" or sign that the aftermath of Cahokia was somehow an especially dramatic declension.[14] When viewed in a larger framework, however, the events that shaped Cahokia's decline were not just an end but a beginning.[15] Considering the nonhuman parts of the story in particular, we can see how the climate events that contributed to the rise and fall of Cahokia also facilitated other changes. This is where bison come into the story, literally.

It is important to emphasize that bison were not complete newcomers to the eastern prairie when they began to appear in the post-Cahokia ecotone region. In Illinois, in the heart of the ecotone, small numbers of bison lived and died more or less throughout the period from the Late Archaic (around 5000 ybp) down through the protohistoric period, around 1400 CE or so. Adapting over millennia to the intermediate space between the woodlands of the east and the grasslands of the west, bison browsed and they grazed depending on conditions.[16] But if they were clearly present for that long precontact Indigenous history, it seems likely that bison were not a big part of human subsistence for most of that time. We do know that archaic people must have hunted bison to some extent. This was one of the exciting conclusions made recently by scientists studying a cache of bison remains near modern-day Peoria. They postulate that one of the animals represented the oldest bison "kill site" in the local archaeological record, from around 2300 ybp.[17] But other evidence from these sites makes clear these animals represent the opportunistic stalking of individual animals, not regular and large-scale herd hunting. Archaeological research suggests the animals' population was small and that they did not constitute a large part of Indigenous subsistence.[18]

Then things changed. Bison remains appear in larger quantities in eastern prairie archaeological sites dated to the 1300s, and then increase significantly in sites from the 1500s forward. Recent systematic attention to the biogeography of the animal makes the pattern unmistakable.[19] Since we now have a clearer sense that the animal's population rose in Cahokia's wake, we are better able to understand why it did.

◆ ◆ ◆

Surely part of the explanation has to do with the animals themselves. Consider a bison born on the Illinois prairie in roughly 1500. She may have been part of a dynamic historical shift, but she was also part of a long species history, the trajectory of which may help us understand what was going on in the eastern prairies at that time. Perhaps the place to start is with some consequential events in the far distant past, forty-five to fifty million years ago, when ancestors of modern bison developed the special equipment needed to pull energy out of cellulose-rich grass forage. This digestive ability, characteristic of all ruminants, actually predates the evolution of the temperate grasses and rests on the evolution of a four-chambered stomach structure in which dense plant matter can be broken down. In this biological adaptation, ruminants did not evolve the ability themselves to produce the cellulase enzymes needed to break down cellulose-based walls of plants and thus release the fatty-acids and carbohydrates contained as energy. Rather, their special stomach structure created a happy home for the *bacteria* that created those enzymes, in effect a kind of internal brewery, where fermentation can radically extend the animals' digestive capabilities. Reflecting the importance of symbiosis and partnership in the evolutionary history of these animals, the ruminants' innovation was to cultivate, nurture, and direct a crucially important set of bacterial partners, a microbiome that would become the key to exploiting one of the major metabolic strategies among the plants of the world: the grasses' C4 carbon pathway.[20]

As discussed earlier, grasses first evolved the ability to outcompete other plants in droughty or hot conditions around 23 million years ago. With the advantage of the rumen now adapted to the predigestion of those plants, bison ancestors soon proliferated in what was now becoming one of the most important vegetation zones of the world, the temperate grasslands. Meanwhile, just as grazers coevolved with the bacteria that enabled them to unlock the calories in what were otherwise inedible grasses, other animals—predators like wolves, lions, saber-toothed cats, and eventually humans—coevolved with the grazers by living parasitically on them in order to tap the energy in grasslands that otherwise might have remained inaccessible. Ruminants became a new center of complex energy transfers, converting the photosynthetic energy of these intensively productive primary producers—the grasses—into the two-thousand-pound bodies of fat and muscle that became a boon to others.[21]

The ancestors of our bison came into this history at a particular moment. One million years ago they broke off from the primitive cow family—*Leptobos*—and transformed. Bison became northern specialists and true open-grasslands specialists, as contrasted with ancestors that favored forest edges. The *Bison priscus*, with its huge body and large horns, came to north America as many as six hundred thousand years ago. *Bison latifrons* evolved in North America around three hundred thousand years ago and moved throughout the increasingly temperate grasslands all over what would become the western United States. Two successors, *Bison antiquus* and *Bison occidentalis*, were short lived, but *antiquus* thrived in the niche left by the megafauna that disappeared around ten thousand years ago from what Dan Flores calls the American Serengeti. Continuing *antiquus*'s success, *Bison bison* evolved around five to ten thousand years ago. In comparison to its ancestors, *B. bison* thrived on a much vaster scale. Indeed, its evolution and rise represents a truly special ecological success story, an ungulate irruption of epic proportions. Shortly after evolving as a separate species, *B. bison* became—in terms of biomass—by far the dominant mammal of the North American grasslands. *B. bison* was a "weed species," a true opportunist whose destiny was to fill empty niches left by Pleistocene megafauna and eventually shape an entire grassland ecology.[22]

In its long phylogenetic history, bison changed shape and size, growing from *priscus* to *latifrons* and then shrinking to the modern species. The collections room at the Illinois State Museum allows one to contemplate the radical changes in the bison's physical *body*—especially in the size of skulls and horns—over its vast history. But even more important are the behavioral changes that evidently came along with the bison's evolution over generations. Bison ancestors had distinctively varying defenses that probably determined their differential fates as they interacted with predators. In particular, *latifrons*'s and *antiquus*'s hooked and forward-oriented horns suggest that those animals stood and fought when threatened by predators. This may have worked for millennia, when the main predators were saber-toothed cats and dire wolves. But when sophisticated human hunters with increasingly effective hunting technologies and social organization began preying on bison ancestors some fifteen thousand years ago, it is not surprising that the humans helped push these bison toward extinction (although, to be fair, *latifrons* was long gone by then). By contrast, *B. bison* adapted a different strategy. They did not stand and fight but instead developed the opposite habit of running away.[23]

Running away was not the only aspect of its evolving behavior that helps explain *Bison bison's* coevolutionary success against predators, especially humans. Even more important was the feature most conspicuous about the animal's behavior: social herding. While *B. latifrons* lived in small groups, *B. antiquus* slowly transformed its smaller, less cohesive aggregations into larger, more hierarchically structured herds. By the time *B. bison* came on the scene, the animals often lived in enormous herds. This is important. As the evolutionary biologist W. D. Hamilton argues, herding was an effective defense mechanism for weak-eyed animals that spent most of their day on flat grasslands with their heads down, grazing.[24] By gathering in herds, they "pooled" their eyes, making it much more likely that a predator would be spotted before it could attack. Another important result of this behavior was that weak animals could be pushed to the edge of the herd, and thus "sacrificed" to save the stronger and dominant individuals at the herd's center. Herding behavior ensured survival for certain individuals and for the group as whole. Surely we'll never know what herding meant to our individual bison born in 1500 in the tallgrass. From our perspective as outsiders, though, these are some of the things it surely meant in history.[25]

The herding behavior of bison is key to our story. Congregating in large herds, bison of the past took advantage of the great expanses of high-quality forage in the biome of the great plains to extend its range in the Holocene. The quality of that forage varied from place to place and in many dimensions. Perhaps most conspicuously, the quality of the grasses changed in a seasonal cycle, resembling what anthropologist Douglas Bamforth once described as a seasonal hourglass.[26] For bison, whose physiology interacted with these cycles in what biologist Dale Lott calls a "fat economy"—a system of boom-and-bust nutrition timed to the annual growth and curing of the grasses—life could frequently be limited by scarcity, especially as herds became larger and larger.[27] To manage the seasonal pulse of scarcity, then, bison herds—especially very large ones—had to be on the move. Migrations could cover long distances, something for which bison were well equipped by evolution.[28]

But if seasonal pulses in forage quality prompted annual circuits across long distances, it was during climate extremes—the same climate events that established and affected the ecotone in the eastern prairies over the Holocene—when bison herds moved especially great distances, from the center of their "baseline" ranges in the northern and southern plains.[29] The principle behind these migrations was rooted in the animals' distinctive survival strategies. While bison graze randomly at the level of specific patches in any given location, they are selective at the level of landscape and will go a long distance in search of new, high-quality forage.[30] In moments of dramatic climate change such as the Hypsithermal or the Scandic Drought (300 to 900 CE), evidence shows that bison herds moved vast distances, even abandoning the entire southern plains for long stretches. Migrations and fluctuations like this shaped plains Indian history in the nineteenth century, complicating the image of supposed "equilibrium" in plains bison cultures of that time.[31]

The fact of long-range bison migration and population fluctuation is crucial for our story. Tallgrass prairie, particularly at the eastern edge, was not always attractive to grazers like bison. The dominant grasses in tallgrass prairie were species like big bluestem which, while nutritious in the early part of the growing season, by late summer contains much more cellulose than grazers might prefer.[32] This made tallgrass forage suboptimal relative to grasslands rich in bunchgrasses and fescues, like those found further west in the Great Plains. In wet periods in particular, bison may well have avoided eastern prairies and preferred the mixed-grass and short-grass prairies further west. Even the small number of bison present in the eastern prairies over the long Holocene seem to have developed not a *grazing* but a *browsing* habit, focusing on woody species and forbs rather than on the local grasses, and probably not in large herds at all.[33] But climate change episodes, especially droughts, were a disruption that could change all of this. Long-term droughts in particular could and did "push" the big western herds of bison off their normal locations by degrading the forage of western bunchgrass. And since the moisture gradient of the continent often made drought conditions less severe further north and east, these "push" forces could bring bison herds in search of forage closer to eastern tallgrass prairies that now appeared relatively attractive, especially since drought conditions possibly *improved* the graze-ability of tallgrass by holding down woody species and certain forb species.[34]

All of this suggests a basic hypothesis about how these features came together in history. During the wet and warm period at the turn of the millennium—the same Neo-Atlantic Episode (850–1250) or Medieval Warm Period that created favorable conditions for maize agriculturalists at the start of the Mississippian period—forage conditions on the plains improved and "grew huge bison herds" all over the plains. For the most part, grazing bison avoided the eastern prairies in this warm, wet period. But then, as we know, things changed. Drought conditions—the Pacific Drought that stressed Cahokia and contributed to its decline—inspired these now booming herds of western bison to move. For hundreds of years, between 1250 and 1500, the animals largely abandoned the completely arid and droughty southern plains for the relatively wetter (if not always wet enough for reliable corn cultivation) areas closer to the eastern prairies.[35] Some herds may well have wound up concentrating in the tallgrass prairies of places like modern Minnesota and southwestern Wisconsin, perhaps even "knocking on the door" of the prairie-forest ecotone of the Illinois Valley. These herds became something like climate refugees, tentatively occupying what had previously been mostly unattractive forage but now provided something like goldilocks conditions for grazing.[36]

While the animals' long phylogenetic history helps us understand the ways in which the climate changes of the medieval Midwest may have triggered a bison migration, pushing and pulling them up to the threshold of the eastern prairies, this background of climate change and animal behavior is not the full story. To understand the experience of our bison born in 1500 in the Illinois Valley, we now must return to human history in the aftermath of Cahokia and rethink the role of people in the coming biological shifts that defined her habitat.

◆ ◆ ◆

By 1500 the human population of the region around Cahokia's previous center—the upper Mississippi Valley south to the Missouri bootheel, the American Bottom in particular, and parts of the lower Ohio Valley—declined dramatically. The population trend is so conspicuous, it has become a kind of historical puzzle.[37] But we should be cautious about how we understand this moment, avoiding tropes of Indigenous decline and disappearance rather than persistence. Beginning in the nineteenth century, generations of American historians and settlers denied that the builders of Cahokia were the ancestors of modern Native Americans, and they invented romantic fictions that drew a sharp disconnect between Cahokia's "lost tribes" and living Indigenous people. The problem with the narrative of what archaeologists have called the "vacant quarter" is that it similarly evokes romantic notions of abandonment, disconnection, and vanishing.[38] We would do better to recognize that population decline in the wake of Cahokia was not total. Despite a smaller population in the region overall, the period witnessed in-migrations as well as a more general "population rearrangement" that affected much of the eastern woodlands.[39] The fact of population decline must not obscure another important truth: Indigenous history continued in the wake of Cahokia. Indeed, owing to the interaction of people and environment in the droughty period beginning in the twelfth century, population decline helped *make* a distinctive history.

Even if the specific area of central Cahokia and its surroundings did "empty out" for a significant period, perhaps because it did, other people and Cahokia descendants rose to new prominence in the period of the Pacific Drought. Most important are the large group of people categorized by archaeologists as the Upper Mississippians, including the Oneotas and Fort Ancient cultures, whose beginnings were contemporaneous with Cahokia's Middle

Mississippians. As Cahokia fell, the lifeways of these two groups became dominant in the region. Smaller in number and much more spread out in smaller settlements than their Mississippian predecessors, these people took over, inventing an important interaction, a culture zone, between the easternmost prairies in places like the Illinois Valley and points just to the west—the prairies of Minnesota and Iowa, which as we have seen were becoming the climate refuges for bison on the plains.[40]

Oneota people most definitely had western connections. The so-called Bold Counselor Oneotas arrived in the Illinois Valley in an "intrusion" that probably originated from the Lake Pepin or Redwing area of modern-day Minnesota. To the north, so-called Fisher phase Oneotas rose in several sites around lower lake Michigan, starting as early as 1100. The signature feature of the Fisher phase group was "fairly large villages and smaller specialized extractive camps," part of a mixed-use lifeway that distinguished them sharply from Mississippians. The Huber phase Oneotas were similar to the Fisher, centered especially in the Sag and Des Plaines Rivers near modern-day Chicago. These people too had western connections and may be ancestors of contact-era Winnebagos or Ho-Chunks.[41]

The Oneotas were marked archaeologically by distinctive lifeways and a material culture that featured characteristic tall ceramic jars. They were agriculturalists and used bottomland fields to grow maize, squash, and other crops, while also exploiting wetland edges in a "broad spectrum" subsistence pattern.[42] Finally they exploited the uplands as well, and probably especially during the droughty conditions that dominated after the Medieval Warm Period. One theory suggests that the Oneotas were a variant of the Mississippian culture that had adapted to the more agriculturally marginal northerly reaches of the Midwest, where farming was less productive than in the Middle Mississippian heartland; it may have been this marginality that constrained these groups and made the Oneotas less sedentary and less politically centralized than their southerly neighbors. But another way of thinking about the Oneotas is that they positively and prosperously adapted themselves to the diversity of a specific niche in the prairie-forest ecotone, relying variously on hunting in the prairie uplands and on fishing and gathering in the wetlands, in addition to maize-based river valley agriculture.[43] If middle Mississippians concentrated their economies on maize, even to the detriment of their nutritional health, the Oneotas had a much more varied subsistence base.[44] This diversity and adaptation to "localized environments"— particularly the hunting component—is what makes the Oneotas so interesting for our story.[45]

Evidence suggests that one of the most important parts of Oneota subsistence centered around grazers in the uplands, including bison. Indeed, it is in the archaeological deposits left by these groups that researchers have traced the slowly rising bison population in the eastern prairies in the wake of Cahokia.[46] As illustrated by the characteristic "end scrapers"—bone tools found at many Oneota village sites and thought to have been used primarily for hide processing—the Oneota groups were avid hunters of large animals. Meanwhile, trash pits associated with these groups feature bison remains, suggesting the importance of the animal in the Oneota diet.[47] The overall pattern of Oneota village organization—the sizable villages combined with smaller seasonal extractive camps—suggests the ways bison hunting probably shaped the lifeway of these people. To be sure, in the context of drought, Oneota groups may have been "forced" into this mode of subsistence by stresses on their agriculture.[48] But their use of newly arriving bison in this period of climate change may also have been a kind of opportunism.

Intriguing clues suggest that Oneota residents of the post-Cahokia ecotone region were in the process of becoming bison people. Overall, the Oneotas' history may represent a "readaptation" to changing conditions, a cultural life organized around an economy divided between upland hunting of bison and elk and localized hunting and mixed agriculture in the lowlands. In an emerging form, here was the lifeway that would later characterize the Indigenous peoples of the contact-era ecotone region, including communal bison hunts and probably long-range travel in pursuit of the animals. To be sure, there is no evidence of bison herds of any large size in the Illinois valley in the 1400s and 1500s, but they likely were present in more western parts of the region where Oneota groups originated and interacted, and herd hunting may well have begun gradually, as the species shift began.[49] Even for easternmost Oneota groups, whose local environment probably did not yet feature large bison herds, intriguing evidence—including in Huber sites near modern-day Chicago—reveal how bison came to feature in ceremonial life. One archaeological cache from this period includes a celt-shaped pipe with bison effigy, a possible "mythical or spirit buffalo," illustrating the relationship between the people and the animals.[50]

Alongside the emerging Oneota bison hunters, a second important culture group joined in this eastern prairie interaction in Cahokia's wake, a subset of the so-called Fort Ancient tradition of the Ohio Valley. Also an "upper Mississippian" culture, the Fort Ancient groups in the Ohio Valley had risen as maize farmers and possible inheritors of certain Hopewell traditions at the same time as Oneota groups were entering the eastern prairies. Sometime after 1450, representatives of the Fort Ancient tradition—or perhaps interlopers among them—moved to the Illinois Valley in another "intrusive" migration from the east.[51] This distinctive subset of the larger culture is traceable by a characteristic pottery tradition known as the Danner series.[52] Like the Oneotas, the Fort Ancient–associated cultures were on the "fringes" of the Middle Mississippian frontier culture. They joined Huber-phase Oneota groups in a new mixed economy, probably also joining in the regional interactions with increasingly bison-based western Oneota groups.[53]

To be sure, it is impossible to distinguish verifiable sequences of cause and effect in this history. But the coincidental development of these post-Cahokia Oneota cultures along with the environmental shifts of the period suggest a hypothesis for how this human history set the stage for a change in the faunal life of the region. First, during the Mississippian and perhaps even Hopewell periods, Indigenous peoples of the Midwest concentrated on maize agriculture and may have avoided hunting, allowing small bison populations to proliferate at the edges of the prairies in the Medieval Warm Period.[54] Then, as Mississippian societies emptied out and drought conditions seized the region, whatever low pressure the Mississippians may once have put on bison in the region was even further eliminated, and bison seeking forage in the context of drought found themselves in a region with few human hunters. In this context the small groups of Oneotas, especially in Minnesota and the La Crosse area of Wisconsin, took advantage of the hunting opportunities presented by increasing bison populations in the context of the agriculture-stressing event of the Pacific Drought. Evidence suggests that some Oneota peoples at this moment were even enticed to a larger-scale bison economy, feeling a "bison pull" toward the plains. The Oneota world dominating the easternmost prairies at this moment was certainly connected to this emerging bison culture, and influenced by it.[55]

In this sequence, any bison herd that ventured east into the Illinois River Valley and far eastern prairies initially may have resembled a kind of "found" resource, its presence a "wild" occurrence shaped by a combination of bison behaviors, the invisible hand of climate change, and the indirectly "anthropogenic" factor of Cahokia's decline. But now another (and much more consequential) climate change took place, the final trigger of species shift that both created new potential for bison in the region and made the new human-bison relationship more intentional and proactive than ever before.

Beginning in the 1500s, climate changed again, turning suddenly cooler and wetter at the start of the global climate event known as the Little Ice Age (LIA). All over North America landscapes came out of the long droughty conditions at the end of the medieval climate anomaly into a much more intense level of productivity. This new climate event was one of the most significant moments of change in the history of Great Plains bison herds, leading to an expansion of unprecedented proportions, perhaps reaching their greatest absolute size in terms of biomass.[56] When a certain number of these animals came "spilling out" of the Great Plains and into the eastern prairies, wet conditions in the eastern prairies probably simultaneously created difficult forage conditions for grazers. Evidence and speculation suggest what happened next. The newly ascendant Oneotas and the Fort Ancient–associated residents, long engaged in a western cultural interaction with neighboring bison people in places like Minnesota and Iowa, began to enhance the eastern prairie forage with fire, in effect cultivating the first large herds of bison in the region. Hardly just a "wild" event, the rise of the bison population in the eastern prairies was in part the work of humans who were purposefully managing the otherwise unattractive prairies to suit their bison partners in a moment of continuing climate change.

◆ ◆ ◆

For our bison born in the tallgrass prairies in the 1500s, times were indeed changing. Where some struggled through the conditions of the Little Ice Age in North America, the period was positively "elysian" on the plains. Plains bison herds grew to somewhere between 25 and 100 million animals, pushing up against the extraordinary carrying capacity of one of the greatest grassland ecosystems of the world. Before long, bison herds began to "overflow" the plains themselves, as population pressure created for bison herds the same kinds of "push forces" that had in the previous dry periods caused the animals to take refuge in the eastern prairies, only now for very different reasons. They were like "toothpaste," Dan Flores writes, squeezed "into every nook and cranny" of the North American grasslands—and even into some woodland edges in the east—as the climate cooled and the Little Ice Age unfolded in the 1500s.[57] One way to think about this is that bison became an embodiment of the flux that was a central characteristic of the ecotone region, a case study in rapid change in response to new conditions. To extend the metaphor of the previous chapter, the proliferating animals went surfing—riding an incoming wave into the prairie-forest tension zone's characteristic shoreline of grass.

Given what we know of the dynamics of that shoreline, the bison's occupation of the tallgrass prairie in the Little Ice Age would not likely have been a straightforward result of climate change alone.[58] For one consequence of the LIA in the flux-prone prairie ecotone in this period was a natural response to increased moisture: a process of reforestation—or forest succession—of the grassland edge. Reversing the drought conditions that likely greeted the region's early post-Cahokia bison pioneers, the moist conditions of the Little Ice Age now brought an underlying tendency for the growth of woody C3 plants in the prairie assemblage. In a flux-prone landscape, this would have been a significant force.[59] In this context, the latent tendency of the vegetation of the region would likely have made the tallgrass inhospitable habitat for large bison herds, especially given the optimal conditions available to them to the west.

Here is where the now-generations-old Oneota presence in the region probably became consequential. In the context of the Little Ice Age, the tallgrass prairies should have been experiencing a process of reforestation. Yet evidence suggests they were not, and instead the grasslands were expanding.[60] Moreover, the primary force holding back reforestation—the force preventing the victory of trees in what Frederic Clements described as the "competition of vegetation types" in the tension zone—was fire. We know this from pollen records: charcoal concentrations in sediment samples from the mid-Mississippi Valley increase around 1500.[61] Though fire was surely a "natural" phenomenon, and present as a factor throughout tallgrass prairie history in what was always to some extent a pyrogenic plant assemblage, it seems improbable that nonanthropogenic fire would have been frequent enough to maintain prairie in the context of a cool wet period like the one that began in the 1500s.[62] In other words, the grazeable grasslands in the eastern prairies into which bison "spilled" were even more an anthropogenic creation than they may normally have been in the Holocene. They were created—and purposefully—by the descendants of the Oneotas and Fort Ancient peoples, whose ascendance in the prairie edge in the wake of Cahokia has too often been ignored. Perhaps both at the level of the individual prairie patch but also at the larger level of the landscape, the Oneotas almost certainly created and maintained significant areas of tallgrass prairie in this period.[63]

Upstreaming—or reading historical accounts from a later period for inference about an earlier period—suggests the ways in which Indigenous peoples likely practiced an anthropogenic fire regime in the 1500s, as bison herds grew in the region. Eyewitness accounts from Louis Hennepin to Jacques Marquette are full of descriptions of what we might call niche construction: systematic prairie burns practiced by Indigenous people in the early contact period, the heart of the LIA. We will explore these practices in greater detail as we learn about the ways in which pedestrian bison hunting worked in the contact era. For now it is important to note how clearly records show a nearly annual pattern of prairie maintenance, one that was focused on increasing the population of bison herds in the region by the Indigenous peoples of the prairie ecotone, particularly in the Illinois Valley.[64] Burning in the fall and often also sometimes in the early spring, Indigenous peoples of the prairie edge were such enthusiastic landscapers that Robert La Salle expressed surprise to encounter a single unburned prairie patch while traveling through the Illinois Valley in the fall of 1680.[65] In the context of the wet conditions of the LIA, fires were surely crucial to the maintenance of the prairie edge and to the grazing bison in the region.

Even at low resolution we can think about this as a kind of domestication, or perhaps as a human-animal partnership. Sharp distinctions between the "domesticated" and the "wild" in our concepts of nonhuman nature can be misleading, and the movement of bison herds into the eastern prairies is definitely a case that require us to rethink simplistic categories.[66] To be sure, the relationship between people and bison here was not "full domestication" in the sense of humans completely controlling the reproduction and conditions of life for their partner animals. Oneota people of the eastern prairies, no less than other people throughout the long history of bison-human relations, were simply not "bison-y" enough for the animals to accept them as part of the herd, much less as leaders of it, as true domestication would require. Bison never became, as Aldo Leopold put it, "servant livestock."[67] Yet neither were bison simply existing in the prairies on their own accord, behaving in the "natural" ways they would have absent their human partners. Indeed, people and bison did embrace each other to an important degree, each performing work for the other, relying on and defining each other. They were an example of that other kind of close human and nonhuman relationship—the commensal.[68] It is no exaggeration to say that people and bison herds *made* each other, together with other entities in this assemblage—grass, climate, fire, and the all-important bacteria inside the grazers' rumen. Managing bison habitat to maximize its grazeability and encourage herds to proliferate, people enticed the nonhuman animals to serve as calorie converters in the eastern prairie grasslands. In turn the bison herds may be said to have encouraged people to satisfy their needs. Neither a case of fully wild behavior nor of domestication, this was something more complicated: a human-animal partnership and kind of "co-constitution" of Indigenous peoples and bison acting together in an important moment of change.[69]

Some of this is speculation; it is impossible to know with precision the particular causal chain that resulted in bison herds arriving in the Illinois Valley in large numbers at the start of the LIA. What we can be sure about, however, is that their appearance was new, embodying a consequential change in the flux-prone bioregion of the prairie-forest tension zone. Soon several conspicuously large villages dotted the ecotone landscape creating places where archaeologists identify some continuity between these Oneotas and Fort Ancient Upper Mississippians and groups from the contact era: the Illinois, the Ho-Chunks, the Meskwakis, the Shawnees, the Eastern Sioux, and others.[70] While the relationship between all these changes in human geography, climate, and animal biogeography probably cannot be known precisely, the evidence seems to suggest something both "natural" and "human"—a kind of world-making, to use the language of multi-species scholarship—was going on here. It was an important entanglement of the ecotone's potential for change with people newly ready to shape that change.

◆ ◆ ◆

To picture the eastern prairie bison today, we need only go to one of several reconstructed prairies in the state of Illinois, where restoration ecologists have lately reintroduced wild bison herds, prairies like Nachusa Grasslands or Midewin National Tallgrass Prairie. There they are in the springtime: the massive animals, heads down and snouts obscured by ground-level vegetation, luxuriating on a sunny day and selectively munching on the shoots and leaves of grasses and forbs. There they go in the later summer: hugging streams for water and an escape from oppressive heat. There they are in the winter: closer to the edge of the woods, taking shelter from cold winds. Watching them moving slowly across these re-wilded landscapes it is easy to be lulled into a sense that these animals and this landscape were made for each other, that the herds' constant grazing is a "natural" phenomenon and reflective of a "stable" or even a "climax" tallgrass prairie ecology. The idea is

Bison and calf grazing at Midewin National Tallgrass Prairie, Will County, Illinois, 2017. Photo courtesy Richard Short/USDA Forest Service's Midewin National Tallgrass Prairie.

in many ways similar to the feeling you get when you stand in the Illinois State Museum's collection room and behold the massive horns of the *Bison latifrons* or the strangely rock-like semi-fossil of a four-thousand-year-old *Bison bison* tooth. Out of context and sequestered from any sense of the human events that accompanied the lives of these creatures in this region, we instinctively imagine them as almost timeless embodiments of one of American culture's most problematic but fascinating constructs: wilderness.

These encounters—with living bison in reconstructed tallgrass prairies or with an ancient natural history collection in an air-conditioned building—can thus be misleading. Far from stable and constant, much evidence proves that the bison in the easternmost prairies were always a kind of contingent presence. What seems clear is that the bison herds that French eyewitnesses described when they arrived in the Illinois Country in the 1600s were not a timeless phenomenon or some straightforward representative of unchanging "nature." Indeed, while grazers were always intermittently a part of many prairie landscapes, these larger herds in the eastern ecotone in the 1600s were definitely a novel phenomenon, an historical event. Arriving in the midst of climate change and ruptures in the local human geography, they were embodiments of the ecotone region's essence: fluctuation. They were intimately tied up with the history of an unsung group of Indigenous people who controlled the tallgrass prairie region in the wake of Cahokia.

As an event, this species shift was not just a curiosity, or a piece of trivia or local color, fit for the obsessions of prairie buffs and natural history aficionados. Instead, it shaped much about the subsequent Indigenous history of this place. Students of early American history should be prepared to accept this premise intuitively. Over the past generation we have learned how species shifts were some of the most significant parts of the early American past, creating great disruptions and opportunities in many ecological contexts, from New England to the South and on through the "frontier" process. Furthermore, many environmental historians have recently elaborated the ways in which human-animal relationships in particular could create conjunctures of great power and change as people and new animal partners built lives together, whether in coevolutionary, domesticated, or commensal relationships. When it comes to bison specifically, we know how much these animals shaped life and power for people in the grasslands, in one case supporting a rise to power so dramatic that it even resulted in "reversed colonialism" on the part of eighteenth- and nineteenth-century Comanches. Together with horses, climate change, and an equestrian cultural revolution, bison on the plains helped Indigenous people create a version of ecological imperialism in reverse.[71]

My argument is not that the bison created the foundation for Indigenous empire-building in the tallgrass prairies of the 1600s per se. Instead, it is that the energy that bison created for people in the tallgrass prairies had more subtle and complex implications. Without bison, the massive store of solar energy embodied by the prairie grasses was locked up in much of the Midwest. In particular, in the future state of Illinois, the heart of the ecotone, prairies represented something like twenty-five million acres of land, fully 55 percent of the landscape.[72] When herds of bison arrived, they made available to people the solar energy contained in some of the most productive grasslands ecosystems in the world. Meanwhile, unlike many grasslands in the West, those prairies were also closely adjacent to several other rich ecological zones—wetlands and forests and alluvial systems. All of this was about to create an historical edge effect, a conjuncture of opportunity in this distinctive place.

To capture the new energy source that the bison and the larger regional mosaic represented, however, the people of the tallgrass—the successors and descendants of the Oneotas and the Fort Ancient associates who became known to history as the Illinois, Ho-Chunks, Myaamias, and others—would have to commit themselves. To put it more specifically, they would have to *work*. Using fire to make the prairies attractive to grazers was just the beginning. Moving forward, they would base their lives around those animals, adapting an ancient kind of large-game hunting to the new superproductive ecological context. Inventing a novel pedestrian bison culture that historians have too often ignored, they would fit their own lives to the material demands of the environment and the animals on whom they depended to harvest its energy. Along with climate, people had almost certainly played an important role in bringing the animals to the prairies. Now the animals in turn would shape the culture of new groups of prairie people.

3

THE RUN-UP

EUROPEANS WHO KNEW THEM IN THE SEVENTEENTH AND EIGHTEENTH centuries knew one thing about the Indigenous inhabitants of the tallgrass prairie ecotone: the prairie people were *runners*. It was a conspicuous fact. Almost all French eyewitnesses who came to the prairie region in the contact era—explorers, priests, imperial officials—noted the prairie Indians' prowess as *coureurs*. Traveling through the Illinois Valley in the early 1680s, Henri de Tonti, who by the time of his expeditions into the midcontinent was familiar with many other Indian groups throughout the Great Lakes, singled the Illinois out. "They are extremely well formed," he wrote. "They are the best runners in America." Robert La Salle, who traveled to the region at the same time as de Tonti, identified the Illinois as "great runners" and "fleet runners." Pierre-Charles de Liette, who spent several years among the Illinois in the late seventeenth century, said that the Illinois "triumph" at running, exploiting this particular talent to build power, "for they know that the enemy cannot run as well as they." Some eyewitnesses commented even more concretely about the running abilities of prairie Indians. Traveling to Illinois Country in the mid-eighteenth century, Diron D'Artaguiette asserted that "I have not seen any Indians who are more agile or who run more swiftly than the Ilinnois," and then included some eye-popping specifics: "I have seen them go . . . a full 10 leagues [25–30 miles], in four hours. This is an absolute fact."[1] Referring to a Meskwaki man in 1731, one anonymous Frenchman observed that the runner could make around seventy miles in a single day.[2]

It's worth pointing out that these were not toss-off descriptions. Of course, it is not uncommon to find European eyewitness accounts in the contact era remarking on Indians' physical prowess, dexterity, and skill of various kinds. Indeed, such descriptions were where Europeans often established important racial tropes, including, for instance, a romantic and exotic discourse around Indian bodies that was as much a rationalization of colonialism and colonial social hierarchy as it was a window into actual biophysical realities.[3] Yet the descriptions of Indigenous prairie runners by Europeans were not just generic and exoticized discourse; French eyewitnesses were singling the prairie people out as categorically *the best* runners they had ever seen. "They all walk faster than we do and are very good runners," wrote Antoine-Denis Raudot, co-intendant of New France in the early 1700s. But "among the nations there are some who are as much better than others as these others are better than we are." Overall, Raudot insisted, the Illinois "run better than any other people."[4]

We should, in other words, take these descriptions seriously. The prairie people of the contact period—the Illinois and Meskwakis and their neighbors—were unusually good at long-distance running. But if this is true, it begs the question: *why*? How should we understand this distinctive and obviously impressive aspect of the prairie peoples' lives? On the one hand, of course, we might think of prairie Indians as part of a longer tradition of Indigenous running, locating the Illinois and their neighbors in a long line of famous Indigenous runners like the Rarámuri, Boston Marathon repeat winner Ellison Brown, Olympian Billy Mills, and numerous Hopi running champions. Indigenous Americans had and have a special connection to distance running across time.[5] But perhaps that is not the whole story. For one thing, even if many Indigenous people are and have been enthusiastic and accomplished runners, we might ask why the prairie people in particular were such exceptional runners compared to other Indians of their time, as eyewitnesses make clear? As we explore this question and consider the particular context of the place and the lifeways in which prairie Indians' running evolved, we might start to wonder whether running was somehow connected to that place and that lifeway.

This possibility becomes especially intriguing if we follow recent scholarship in environmental history and related disciplines to rethink the relationship between human "culture" and "nature," or between the human and the nonhuman. Sure, it is possible on one hand to understand running as simply a *human practice*, a set of competencies and skills motivated by human decisions and behaviors and explained best by reference to certain preferences and ideologies (which is to say "culture" or "the social").[6] Alternatively, it might be possible to think about running ability in the frame of strictly *human biology*, exploring matters like physiology and genetics, as some scholars have done especially in the case of Indigenous runners, although not specifically those from the midcontinent.[7] Both of these approaches put the focus strictly on the human agent in the scenario and categorically divide the realms of culture and biology. But following scholars in recent generations who have taken a so-called ontological turn to explore the ways in which human culture is importantly co-constituted with nonhuman material entities, perhaps it would be better to understand prairie peoples' running as at least partly a consequence of the interaction between the humans and the nonhuman fellow "actors" in their environment. Moving beyond enlightenment dualisms between nature and culture and toward a reckoning with the true hybridity of human and nonhuman entanglements, prairie peoples' running might best be explained in terms of the relationality of people and the environment—the specific landscape, specific ecology, special animals, and a complex and idiosyncratic mixing of energies that took shape in this place and time. In addition to being something more than just another instance of a transhistorical Indigenous cultural tradition, or the consequence of genetic endowments, running for these prairie Indians of the contact era might well have been more than just a human practice to begin with.[8] In particular, their running may have had more to do with the prairie landscape of the early modern period and their partner in its exploitation: the bison.

Sometime in the 1600s environment and human culture interacted and shaped one another in the ecotone, as people created a special lifeway as *pedestrian* bison hunters. Bison were "energy on the hoof," but in a way that depended on particular kinds of human labor as well as on the interactions of the grasses, the seasons, herd behavior, and growth patterns. These animals became the heart of a material system of energy for generating human power and culture. But, like longhorn cattle, sheep, horses, and other animals whose power to shape distinctive human cultural practices may be more familiar in our historical imaginations, bison were most certainly agents in shaping the way people lived. Many scholars have investigated the relationship between people and nature in the tallgrass prairies to assess whether and how the ecology of the prairie was anthropogenic, a consequence of human culture. This chapter looks at the situation in reverse, following historian Timothy LeCain's insight that seemingly wholly anthropogenic and "human" matters make little sense, "unless we first recognize the primal creative and destructive powers of the animals, molecules, and other material things from which they had emerged in the first place."[9] Moving past old ways of understanding historical agency, the point is to explore how human history was shaped in relation to nonhuman entities in the prairie, and vice versa.

And in this story of human and nonhuman entanglement, running was a central part of the relationship, a practice that allowed people to take advantage of a particular energy regime in the tallgrass prairies, as bison arrived there in the wake of Cahokia. Indeed, running—and the pedestrian bison-hunting lifeway more broadly—is a perfect manifestation of how cultural practice is co-constituted with nonhuman environmental realities. Although so often distorted by Europeans as a deficient mode of production, pedestrian bison hunting was indeed a cornerstone of a powerful energy system. But it also shaped the rhythms of human life in many dimensions—running after bison was no simple matter. The practice committed the prairie people to certain techniques and strategies in the hunt, of course, but also more subtly to certain kinds of communal ideals and practices that continued even when the hunt was over. Saying this is consequential. It means that the landscape and the requirements it placed on people helped to create peoples' subjectivities and social relationships as they mingled their work with it in thoroughly embodied ways. In other words, here was a special biophysical assemblage co-constituted with human culture: *a collectif.*[10]

The co-constitution of nature and culture in this scenario is most visible on two levels, or what we might call the "hardware" and "software" of tallgrass pedestrian bison hunting.[11] First, bison created a rhythm and communal pattern in the lives of pedestrian bison hunters, inspiring long-range movement and an annual pulse as large groups of people and herds broke up, came together, and moved around each year. Relatedly, on a second level, bison and people together shaped new individual and collective behaviors and, we might say, *mentalité.* The most important features flowed from the facts of the herd. Bison did not accept domestication, but they did make a place for people in the herd's life, and this in turn created certain imperatives for people to follow as they stalked the animals, killed them, and used them—always necessarily on a large scale. Without totalizing or essentializing or succumbing to determinism, this chapter argues that the large-scale requirements—and affordances—of pedestrian bison hunting were so fundamental as to create the dynamics of human relationships, solidarities, subjectivities, and social patterns.[12] Central to these processes, it turns out, was running, or the act of large groups of people covering long distances in pursuit of animals that could—rather suddenly in the 1600s—help people channel the sun's energy in the tallgrass ecotone into usable calories and human power. Making lives around these animals, working together with them, newcomer groups like the Illinois and the Meskwakis, along with long-established ecotone residents like the Ho-Chunks, became prosperous and created new cultures. Even so, while their labor made them well-off, it also committed them to communal practices that would prove to be—like the landscape itself—unstable and tenuous. To understand all this, let's start with the hunt.

◆ ◆ ◆

To reconstruct the pedestrian hunting lifeway that prairie Indians developed as the herds expanded during the Little Ice Age, we can rely again on a method called upstreaming: using eyewitness accounts and other evidence from after the contact period to look backward. One such account, a true and somewhat unsung classic of early American travel literature preserved today in the collections of the Newberry Library in Chicago, was left by a Frenchman called Pierre Charles de Liette, who lived in a French settlement called Fort Saint Louis near modern-day Starved Rock toward the end of the seventeenth century. In June and July 1688, Liette went on a summertime journey with the Illinois Indians through the Illinois Valley. And while he learned about and reported on countless particulars, both about the Indigenous Illinois people and the distinctive landscape of the tallgrass prairie bioregion, for our purposes the real value of Liette's account lies in his sensitive description of one of the most amazing things he witnessed: the Illinois' pedestrian bison hunt.[13] While only a few generations old, the practice of herd hunting in the eastern prairies was by the 1680s totally consummate, an extraordinary economic practice in early America.

It stood to reason that Liette would devote a good deal of his account to the bison hunt, in part because bison were surely the most stunning and remarkable animals in the region and the subject of much curiosity among Frenchmen. By the 1680s, the Illinois Valley where Liette and his hosts traveled had become roughly the easternmost part of the range of large bison herds in North America.[14] French eyewitnesses noted frequently how these herds dominated the vista, appearing "innumerable" and stretching "as far as the eye could see"—the most charismatic thing about the ecology of the grasslands.[15] Liette knew his readers would be interested to learn about the "great herds" that his party encountered and the manner in which the Indians hunted them. So he spent many pages on an explanation of the hunters' technique. His account was not unique; many others wrote about this as well.[16] But since he participated in the hunt, Liette's account is exceptional.

Obviously the most conspicuous thing about the prairie peoples' way of hunting was that it was done on foot. Totally different for that reason from later and more celebrated equestrian hunting cultures, this "community hunting" was a remarkable practice.[17] Liette noted the way in which his Illinois hosts set out across the prairie landscape "in two bands, running always at a trot," then bursting into "full speed." From other sources we know that a hunt like this could involve as many as four teams and it often took a whole day or even several days; hunters selected the kill zone in advance and started their run from a considerable distance away in order to be able to coordinate the animals' movements precisely toward the spot.[18] Since bison can run over 30 miles per hour at top speed, they can easily elude human hunters for short distances; success rested on chasing them over long distances and thus tiring them out. As Liette noted, this strategy took a lot of organization to be successful—the hunters had to cooperate carefully for the purpose of corralling the animals over a large space. He noted how they enforced the requirement of stealth, appointing "guards . . . [to] prevent anyone from separating from the band and going off alone."[19] Occasionally the prairie Indians would use fire to make a barrier in their efforts to corral the animals. When the bison finally were led to the appointed spot, however, the whirlwind and the carnage began. Here's Liette: "When they were about a quarter of a league from the animals [between a half a mile and a full mile away], they all ran at full speed, and when within gunshot they fired several volleys and shot off an extraordinary number of arrows. A great number of buffalos remained on the ground, and they pursued the rest in such manner that they were driven toward us. Our old men butchered these."[20]

It must have been an incredible sight, not just because of the craft involved in the hunt but also because of the sheer intensity and riskiness of the enterprise. Reflecting on this, in a climax moment of the hunt Liette's account became somewhat emotional. He revealed the intense feelings that accompanied approaching these animals up close, and included what is a pretty rare expression of vulnerability in explorers' accounts from early America: a contemplation of fear that concluded with a moment of unusual self-deprecation. Here's how he summed it up: "[The animals'] appearance filled me with terror, and I withdrew from our troop when I [saw] them approach; which set all the [Indians] laughing, at which I was not a little mortified. It is certain that those animals are frightful looking and usually terrify people who have never seen them."[21] Indeed, Liette's embarrassment continued when he tried to "redeem" himself, only to kill a tiny bison calf which had lost its mother and lay defenseless in the grass. Despite the Indian hunters' assessment that the pathetic and helpless animal "was not worth the shot," the Illinois nevertheless graciously celebrated this minor success by their guest. Liette meekly narrated how his Indian hosts cooked and distributed the meager amount of meat from this animal to more than 120 people along for the outing, "honoring" him just as they would have any young man after a first kill. As an affective experience, the upshot of Liette's journey included many emotions—danger, exhilaration, vulnerability, fear, and, when it was over, satisfaction, relief, and community.

Several hundred years later, readers of this account might be amused by the anecdote of Liette's hunting "success" and the irony of a colonizer put in his place by his Indigenous hosts. But our more powerful reaction to Liette's account of the hunt might be to share his sense of awe—even astonishment—about what he witnessed. Much less famous than Great Plains equestrian bison hunting, the practice of pedestrian bison hunting that Liette described seems to eclipse that other mode for its sheer improbability. While nobody would call bison hunting on horseback a leisurely picnic, at least in that discipline horses did the work of locomotion as well as provided minimal security in case a rapid escape proved necessary. By comparison, Liette's account reveals the added physical work involved—as well as the danger—when hunters simply *ran after* bison. Meanwhile, Liette's account also forces us to consider the limitations of the landscape in which these Illinois hunters were operating. To the extent that many of us know anything at all about the way pedestrian bison hunters conducted their craft, we probably picture them driving animals to dune traps, cliffs, or "jumps" in order to kill large numbers of animals relatively quickly, and from a distance.[22] But the tallgrass prairie region was a flat land conspicuously lacking cliffs; Indian hunters

Pedestrian bison hunt. From Antoine-Simon Le Page du Pratz, *Histoire de Louisiane* (Paris, 1758). Image courtesy Rare Book & Manuscript Library, University of Illinois at Urbana-Champaign.

corralled the animals to a kill zone only by surrounding them in large groups or by using fire, with both techniques surely increasing the danger of the scenario considerably.[23] And if all that is not impressive enough, consider finally that this was all taking place in a *tallgrass* prairie, a vegetation assemblage comprised of wildflowers and grass species like the very appropriately named big bluestem. Anybody who has been to a prairie in June can understand how difficult and intimidating it would be to try to chase a large herd of huge animals through plants that could easily reach eye level of a large adult. The challenges involved in this practice are staggering.

Finally, there was the scale of the enterprise. These hunters were taking animals on a massive scale. Liette reported that on one single day of the hunt the Illinois killed 120 of the beasts.[24] Over the course of their five weeks in the field they took over 1,200 bison in total. Even if these were all cows, and thus on the small side for the species, 120 animals—a single day's kill—would yield, at a minimum, around twenty-seven thousand pounds of useful meat, and as much as forty-eight thousand pounds.[25] If any of them were bulls, as is likely, they would yield even more. Processing this meat was a massive undertaking. To preserve the animals the Indians occasionally salted them, but more frequently cured them into a kind of jerky. Smoking the animals' flesh on frames above a fuel of reeds and aromatic wood that they gathered in woodland and wetlands, they worked on an almost industrial scale over the course of several days. Then they transported the meat. As Liette told it, each able-bodied member of the group carried as many as eight huge flat ribs on their back, hauling them as far as 20 leagues (40 miles) back to the village in the Illinois Valley.[26] As we know from other accounts, dogs, sleds, and sometimes (rarely) canoes helped in this part of the work. Regardless, it is clear that transporting the processed meat was among the most laborious parts of the hunt.[27] Overall, even if Liette exaggerated or got certain details in his description wrong, the bison hunt he witnessed featured an astounding amount of labor, ingenuity, and industry.

Liette's description is thus a valuable window into an amazing practice, and it joins several other French eyewitness accounts that provide an understanding of the techniques involved in what is an often-ignored human-animal lifeway of Indigenous America.[28] Judging by its scale and finely tuned operation, it is clear that by 1680 this was a fully fledged system. The prairie people were not just casually living off the land nor just waiting for the bounty of thousands of animals to come to them. Instead they were performing an extraordinary amount of work, precisely coordinated and purposeful, to maximize this resource.

As we contemplate the nature of that work and of the experience in Liette's account, however, it is clear that something special was going on here. Bison hunting was never a straightforward case of human mastery, nor a scenario in which humans simply *acted on* a passive nonhuman world. Instead, many aspects of the hunt—including even the seemingly human moments like when the Illinois paid tribute to the newcomer Liette with a feast—force us to reconsider a fundamental question: who—or what—is the actor here? As Paul Robbins has argued about the phenomenon of the suburban lawn, environments and nonhuman entities exercise agency to shape our human decisions and our very subjectivities, creating the imperatives and logics that drive our behavior by tangling together "nature," the social, and the cultural.[29] Considering not just the hunt itself but all the ways in which the hunt encouraged or even required people to live in certain ways,

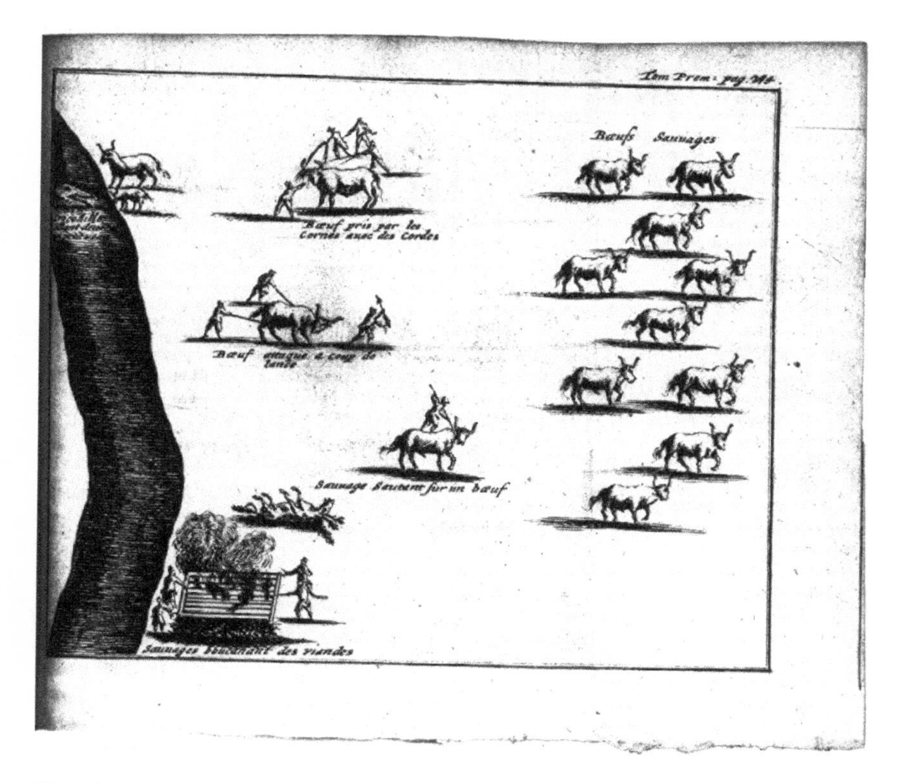

Bison hunt. From Louis Armand de Lom d'Arce Lahontan, *Nouveaux voyages . . . dans l'Amérique septentrionale* (1709). Image courtesy Rare Book & Manuscript Library, University of Illinois at Urbana-Champaign

following the bison was just this kind of two-way street. This is significant. As environmental historian Richard White has influentially argued, people know and shape nature through our work.[30] And we are shaped by nature because nonhuman entities—including animals—perform work too.[31] In the most instrumental sense, the work of a bison was to turn the C4 grasses of the tallgrass prairies into a 2,000-pound body.[32] But in order to access all that energy, people had to shape their lives and organize *their* bodies around the animals. Perhaps most fundamentally, they had to adopt certain community dynamics. Hunting deer—and even hunting bison on a horse—might be something a single person could effectively do. But running down, corralling, killing, and processing a herd of bison on a flat prairie would be inconceivable without a large number of people, a careful division of labor, a great deal of social solidarity and cooperation, and an annual rhythm that structured it all. In a sense it was the bison that made these social dynamics and shaped the prairie people in certain specific and embodied ways. To understand this in depth let us consider first what we might call the prairie peoples' economy *without* bison: the richly diverse seasonal economy into which bison introduced new and different dynamics.

◆ ◆ ◆

It is important to note that bison hunting in the eastern prairies was just one part of a large and diverse seasonal economy featuring resources from different parts of the edgy prairie mosaic. It was not seasonality, per se, that made bison hunting distinctive, but rather the way in which the practice imposed a singular seasonal rhythm on the entire community, all at once and on an unprecedented scale. To understand why this was both unusual and important, we need an overview of the rest of the prairie peoples' diverse economy to which bison hunting was always supplementary and a secondary food cycle.[33] In the wake of Cahokia, the Oneotas and proto-Algonquian peoples who reoccupied the region of the tallgrass ecotone developed a sedentary foraging and agricultural economy that featured, like the lifeways of many precontact Indigenous groups east of the Mississippi, an annual cyclical pattern of great variety, a sequence of seasonal harvesting in wetlands, prairies, and forests. Most of these practices were compatible with small-scale village organizations that typified the groups that reoccupied the ecotone region in the wake of Cahokia. But *not* bison hunting.

Reflecting how lifeways rested on exploiting distinct plant communities in their region, the Miami-Illinois language named five distinct ecological zones.[34] First were the lowland river valleys, where the prairie people located their summertime villages and where they practiced their main agricultural component of subsistence. In these bottomlands, speakers of the Illinois language raised four kinds of maize, or *miincipi*.[35] They also planted *eemihkwaani* (or wild pumpkin squash), *aleciimina* (peas), *iihkihtaminki* (melons), *kociihsa* (beans), *wiipinkwamina* (huckleberries), and lima beans. Raising crops in the lowlands made sense, first of all because soil moisture and fertility were extremely high. At the same time, clearing the land was much easier in these lowlands than it would have been in the uplands and prairies.[36] Farming by the Oneotas and other occupants of the Illinois Valley in the precontact period was a central part of subsistence, but it was small-scale and supplemented by many other annual food cycles and resources.

Occupying the same lowland zone was the *wiihkweehkiwi*, or floodplain. Here, the prairie people took advantage of abundant wild resources, especially for food production. In these areas were abundant summertime berries, such as *makiinkweemina* (blackberries), *napaleeteemina* (raspberries), *meenkaalakiinkweemiša* (dewberries), *mihtekwaapimiši* (mulberries), *wiikooloomphsa* (elderberries), *wiipinkwamini* (blueberries), *akaayomišaahkwi* (gooseberries), *ateehimini* (strawberries), *mahkomiši* (sumac berries), and currants.[37] Certain hardwood and softwood trees were present in the floodplains, including cottonwood, silver maple, elm, and ash, all useful for technology. Of course, in the floodplain were also abundant backwater lakes, where the prairie people could harvest fish and mussels, especially in the lean early spring season.

More remote from the floodplain were the marsh and wetland areas, where prairie people hunted birds and gathered diverse plant resources for food and technology. In the wetlands the prairie Indians harvested the reeds (most likely softstem bulrush, or *alansooni*) to make *apacois*, the main exterior material of their houses. Meanwhile, women scoured the wetlands for roots, in particular the *mahkohpina*, or macopine, the white water lily. French observers were astounded at the abundant resources that the Illinois gathered from the wetlands, as when Liette described the harvesting of what was probably hollow root. Also in the wetlands were "bear's root" or Indian turnip (Arisema tripyllum) and several other species of edible plants. Importantly, the wetlands possessed a high density of useful plants per area of land cover and were the source of a great number of fowl, including cranes, herons, "swans, bustards, wild geese, and ducks of all kinds," which the prairie Indians hunted.[38]

Upland forests, the *mihtekwaahki/mihtehki/ahtawaanahki*, were the most significant ecozone for many subsistence practices of the people of the forest-prairie ecotone. The upland forests contained by far the greatest number of useful plant resources for the eastern prairie Indians, including large numbers of food species (mostly fruit and nut trees), technological and medicinal plants, and habitat for significant numbers of woodland game species—particularly white-tailed deer and bear—all of which were important to diet and technology.[39] In contrast to ecozones like the prairies and the wetlands, the density of useful species in the forest was relatively low; here resources were more spread out, requiring a large amount of forest land cover to satisfy all the needs of the people. Although many of the Indigenous prairie peoples' well-tested ecological traditions were based in the forest, the landscape in this particular ecozone of Illinois Country was relatively poor.[40] In certain moments—like toward the end of the seventeenth century—this built-in scarcity in the regional mosaic could become a problem.

Finally, there were the prairies, *mahkoteewi*. Like the wetlands, prairies had a high density of useful plant species per area of land cover. According to Liette, Illinois hunting parties could support themselves from day-to-day simply by gathering plants as they traversed the prairies in summertime. As he put it, the prairie contained an "abundance of all things." Prairie plants gathered for food included *waapinkopakahki* (goosefoot), plantains, docks, and dandelions. Illinois women also gathered Jerusalem artichokes, wild onions, and other food plants in the prairie.[41] The prairies were also the source for milkweed (*Asclepias syriaca*), which was an important plant for both food and cordage. But the most important resource in the prairies after 1600 was of course the herds of animals, especially bison. By burning the prairies annually or semiannually, the prairie people enhanced the graze-ability of the grasslands and helped promote this segment of the region's ecology, turning the prairies into a giant game reserve. Over time the rhythm of bison production and exploitation became the heart of the subsistence system, the centerpiece around which the rest of the diverse foraging and agricultural system worked.

But if the bison came to drive the Illinois' economy in important ways, in return the animals definitely made different demands relative to other components of the foraging and farming economy that prairie peoples practiced in this edgy mosaic. As a revisionist literature has recently shown in great detail, sedentary foraging lifeways—especially as practiced in wetland ecotone settings common in the Midwest's river valleys—were some of the most productive, efficient, and resilient economies ever devised, and were surely "superior" to concentrated grain-based agricultural systems in the amount of energy they yielded in comparison to the work they demanded from people. Moreover, given the impossibility of controlling or enclosing particular resources or energy streams in them, in many cases they also fostered little social differentiation or political hierarchy. For political scientist James C. Scott and others, this is surely why so many sedentary foragers remained committed to these economic systems for thousands of years, even after the domestication of crops and animals that later became the hallmark of intensive agriculture and state formation. More attractive—because of the great variety and resilience it offered its practitioners, as well as the egalitarianism it fostered—the system of sedentary ecotone foraging featured a simple logic:

exploit the seasonal pulses of the landscape, even while staying in one place. As Scott summarizes, the advantages of the lifeway came from abundance, proximity, and seasonality: "Instead of the population having to shift camp from one ecological zone to another, it could stay in the same place while, as it were, the different habitats came to them."[42]

Understanding how the sedentary foraging economy went hand-in-hand with small-village social arrangements, low-intensity labor regimes, and a decentralized and egalitarian political pattern makes it easier to see how the bison economy introduced a different dynamic. For bison hunting—especially as practiced by the pedestrian bison hunters of the tallgrass prairies—was simply not possible as a solitary or small-group activity, and it definitely required specific kinds of intense work for the people who practiced it. Pedestrian bison hunting created a unique process of large-group consolidation that involved a great amount of cooperation and division of labor—in other words, complexity without great hierarchy.[43] Adapting the concepts of anthropologists Russell Barsh and Chantelle Marlor, we can think of this first in terms of the "hardware" of bison hunting—seasonality and size.[44] In a way we might say it was the specific herd behavior of the animals that demanded and ultimately created a distinctive kind of herd behavior among prairie bison people.

◆ ◆ ◆

The most important aspect of what we might call the "hardware" of prairie bison hunting was the peoples' seasonal movement out of their villages. The bison migration fundamentally defined and controlled prairie Indians' lives, giving them their most important annual pulse: two hunts that punctuated the calendar. Each year, groups in the prairie ecotone abandoned their main lowland villages in "autumn, after they have gathered the harvest," and went to winter hunting camps located variously in the uplands, where they stayed until March. With the arrival of spring they moved back to their main villages in the river valleys to fish and begin springtime planting. As the spring progressed and seeds were sown, members of the village departed for the second main hunt of the year; as Nicholas Perrot notes, "as soon as [planting] is done, they go hunting again, and do not return until the month of July, which is the time when the rutting season of the buffalo begins." If the guiding principle of most every other component of the prairie peoples' subsistence was to stay in one place and let the "different habitats come to them" (as Scott puts it), the role of bison was the important exception. For, as Hennepin put it, "These beasts . . . change their Country according to the seasons."[45]

While of course all of the rest of the prairie peoples' subsistence—from fishing to foraging to farming—created various kinds of seasonal rhythms throughout the year, nothing else required long-distance movement, relocation, and discontinuity on such a scale. Scale is the important issue here. What was regular about these hunts was their timing, year after year. What was irregular about them was their location. Bison could be nearby or far away, and their movements, though partially controlled by the planned fire schedule engineered to control the grazing landscape, were also in some ways probably close to random.[46] Hunters had to follow. As revealed by Jesuit priest Gabriel Marest, who accompanied Illinois hunters in the early 1700s, the prairie Indians went anywhere and everywhere in search of the animals, and often spent the bulk of their hunting season traveling. As he summarized, in writing for an audience of fellow Jesuits who were in general no strangers to difficult travel, the prairie peoples' journeys were distinctively demanding: "These journeys which we are compelled to take from time to time . . . to follow the [Indians] . . . are extremely difficult." The Miami-Illinois word *nina8inepa8i* captures the essence of this lifeway: "I am going to the hunt far away and for a long time."[47]

If demanding and long-range seasonal movements made the prairie Indians' travels distinctive, consider also another dimension of their scale, namely, the size of the groups that undertook such travels. Chasing deer, woodlands caribou, beavers, and other nonherding animals, most hunters of the woodlands regions of the Great Lakes set out in relatively small wintertime parties so as better to stalk and capture solitary animals. For bison hunters the method was different. "When the seed is sown," Claude Dablon wrote of prairie groups, "*all go together* to hunt the wild cattle, which supply them with food" (emphasis added).[48] To be sure, it made sense for prairie people to spread out across the landscape into multiple small settlements—"scattering," as the Jesuits often put it, or "separat[ing] into many bands."[49] But, unlike other groups, these hunting camps clearly remained in contact with each other and coordinated their actions. The reason is obvious. As soon as the larger herds of bison were located, the job of hunting them required massive numbers. As Hennepin puts it, "When the [Indians] discover a great number of those beasts together, they . . . assemble their whole Tribe to encompass the Bulls." Perrot was even more explicit: "The people of an entire village go together to this hunting, and, if there are not enough of them, they unite with those of another village . . . [so] they may be able to drive in a greater number of animals."[50]

Given their coordination during the hunt, prairie people seem to have returned in the summertime to what were in fact relatively large main agricultural settlements. This is an important fact and may reflect a kind of feedback loop in the bison "hardware." Large coordinated social groups made bison hunting more effective. But then in turn, bison hunting—by yielding tremendous quantities of calories and useful material (hundreds of animals' worth for each outing)—supported and perhaps encouraged ever-larger groups to live together. To put it slightly differently, bison hunting required and then also *enabled* the larger populations, and thus in general promoted a social consolidation, especially when compared to other aspects of the subsistence patterns of post-Cahokia Indians of the forest-prairie ecotone.[51] As time went by and especially in the late seventeenth century, these large villages came to have an important relationship to military affairs and the strategic incentives and imperatives of life in a volatile borderland. In the beginning, though, these side effects of following bison were probably experienced as a matter of opportunity, not an act of necessity. Moreover, unlike plains equestrian hunters who effectively abandoned the "safety nets" of river valley agriculture and gathering for a thoroughly nomadic and specialized economy beginning in the 1700s, the pedestrian hunters probably experienced less discontinuity. Instead, they added and adapted bison hunting, large villages, and the accompanying changes to previous social patterns.[52]

Recognizing the relationship between bison hunting and large human population configurations does force us to confront an important tension, especially with regard to how these large populations interacted with the dynamics of the bison herds. This is because bison are k-selected (as opposed to r-selected) ungulates, meaning that from year to year they exhibit patterns of fairly stable populations, slow dispersal, and regular and slow reproduction relative to their "standing crop"—the baseline numbers in herds. The result is a steady and predictable population from season to season, even though extremely variable over longer timescales. This is in contrast to r-type populations (like deer, for instance) that tend to "irrupt" in great numbers at times of favorable forage but then overstretch their available resources only to crash in numbers when conditions change or when populations exceed carrying capacity. The bison's relatively slow and steady reproduction (18 percent growth per year) and slow recovery from major population losses were a significant fact once its presence was established in the eastern prairies. In the cases of the Illinois, the Wabash, and the Wisconsin River Valleys, the herds probably became well-established and their populations somewhat regular year over year.[53]

With ever-larger villages and more sophisticated hunting techniques, however, people could easily threaten the slow and steady k-type population dynamics of bison in the tallgrass. The fact that this seems not to have happened makes obvious the intentionality that governed the way the prairie people hunted. As Louis Hennepin pointed out, for example, the Illinois only killed bulls during their summer hunt, probably so as to protect mothers and calves and ensure the herd's reproduction. Overall, even had they taken as many as "six score" (120) animals on a single day of a bison hunt, the prairie people obviously cooperated to limit their impact on the herds to ensure a continuous supply and to make sure the yield was below the 18 percent threshold that represented a steady herd reproduction level. While Hennepin wrote that the prairie people "were never able to destroy these Beasts [bison]" owing to their steady reproduction, both Hennepin's understanding of ecology and his implication that prairie bison hunters lacked sophistication or efficacy in their technique were mistaken.[54] Instead, the key was their intentional conservation, just one aspect of an extensive set of principles, ethics, and norms that went along with the practice of following bison. These norms, what we might consider the "software" of pedestrian bison hunting, featured a communitarian and cooperative ethic and *esprit* which, just like the seasonality of life, were co-constituted with the animals themselves.

◆ ◆ ◆

Somewhere in the eastern prairies of modern-day southern Minnesota in the year 1680, the Mdewakanton Sioux (or, as Hennepin called them, the Issati), prepared for their wintertime pedestrian bison hunt. As the chosen day arrived, all was business—hunters arranged their weapons and women began assembling the tools for processing the meat. Gathering together, hundreds of people took their places in the growing camp and began discussing how they would approach the task ahead. Yet, as the day for going into the field grew closer, several matters still needed careful attention. First, a chief at the head of the now enormous hunting village sent "four of five of their most expert hunters" to perform a special ritual in the field, a dance made in tribute to the bison and featuring "as many ceremonies as amongst the Nations to which they are wont to send Embassies."[55] Then, when the dancers returned, the rest of the village began an extravagant kind of performance, uniting in a solemn procession featuring "two hundred Hunters" who began "to lament bitterly the Death of those Bulls they hop'd to kill." Next "two of their nimblest Hunters" went out of the village to scout the herds, returning to report their location to "the most ancient of the Company." Then came the climax, the ceremony of the Calumet: "Afterwards they made a Fire of Bulls Dung dry'd in the Sun, and with this Fire they lighted their Pipes or Calumets, to smoak the two Hunters which had been Sent to make the Discovery. Presently after this Ceremony was over, a hundred Men went on one side . . . and a hundred on the other, to encompass the Bulls, which were in great numbers."

There is much we could say about this ceremonialism, what in the Miami-Illinois language might have been known as *irenansecana* (bison dance).[56] For our purposes, what makes this performance fascinating is the window it offers onto the imaginative universe that surrounded bison hunting for the prairie people.[57] What seems obvious is that the entire ceremony was an effort to define and express *community*—community among the people and community between the people and the animals themselves. From this and other episodes, together with a relatively large body of primary sources and insights from Indigenous thought, we can piece together how these ideas of community rested at the heart of pedestrian bison hunting, amounting to a distinctive *mentalité* that ran through and defined these eastern prairie peoples' lives.

As the Mdewakanton ceremony suggests, the first thing to note about bison hunters' ideas of community is that they are inclusive of both people *and* the bison themselves. When analyzing the worldview of pedestrian bison hunters, this is at once the most obvious fact but also quite difficult to fully summarize. Bison, to the prairie people, were no objects, but rather were active and animate subjects. Of course this did not make the status of bison unique in the worldview of Indigenous North Americans; the starting premise of the cosmology of many Indigenous people was and is that the nonhuman world is full of sentient and intentional *other-than-human-persons*, beings to whom humans are associated not in a hierarchical relationship of superiority but rather in a horizontal relationship of fellowship, equality, and reciprocity.[58] In this cosmological frame, animals are relatives, and *manitou* is the animating force that governs us all.[59] People must respect this non-human power and especially in the animals on whom they depend for life. The obligation for reciprocity includes the possibility of sanction. For instance, the Peorias' winter tales—spiritually important narratives kept by oral tradition—frequently revolve around animals who thwart people for failing to respect them or to create an agreement of mutual benefit. At the same time, animals willingly give themselves up as long as the people treat them in a respectful way.[60]

In prairie peoples' cosmology, bison were clearly on a special plane, and their deaths from human hunting were measured as an explicit sacrifice performed for human survival. Since they willingly gave themselves to satisfy human needs, bison definitely had special efficacy within the animal world. Jesuit Jean Mermet learned this firsthand while trying to explain Christian cosmology to a group of Illinois speakers in 1712. As he laid out his theology, the priest found he simply could not convince his audience that a Christian God, because resembling humans, was somehow superior to the spirit of a bison.[61] The Illinois speakers totally rejected this ontological hierarchy. Reflecting the spiritual power of the bison, prairie people also frequently dreamed of bison manitou who could reveal and predict the outcome of human affairs.[62] Perhaps most reflective of the way prairie people considered bison as efficacious persons, the five Mdewakanton hunters who "danced" on the first day of the prehunt ceremonialism did so using the Calumet in order to "to make [an] Alliance" with the animals. In human affairs the Calumet was a technology for community, for making relationships with strangers and for honoring and turning outsiders into friends (as will be seen in chapter 4). By conducting this interspecies diplomacy, and by "lament[ing] bitterly the Death of those Bulls they hop'd to kill," the Mdewakantons expressed their mutual personhood—their community—with the bison. This echoes an instance in which one Illinois hunter performed a ceremony after killing a bison, warning Frenchmen not to approach until the hunter had properly ascertained and appeased the animal's spirit.[63] While modern science might think of each bison death from hunting (or predation in general) in terms of a quite depersonalized "sacrifice"—a function of evolution in which weaker members of a herd are surrendered to protect dominant individuals and the reproduction of the overall herd—for Indigenous hunters the sacrifice was considered in much more personal terms. It was an intentional act by the animal. Bison hunting was not just a matter of technique; rather, it fundamentally involved getting in community with the animals.

If this sense of community among human and nonhuman persons is perhaps inaccessible to nonparticipants in the prairie lifeway, an easier thing for us to analyze and intuit is the ethic of community that bison hunting created among the humans themselves. Bison, like other animals, were tricksters, and chasing them required all the intelligence, organization, and discipline that humans could muster.[64] Thus pedestrian bison hunting demanded a sense of community among the people, manifest in specific rules and norms. The community ethic operated on two levels. First was the imperative of cooperation between groups of prairie hunters. This is straightforward. Considering the way that bison herds defended themselves—by running, often far away, from danger—it was easy for one group to spoil the hunting of other groups either by mistake or intention. Bison hunters thus had a simple maxim to stay out of each other's way, and "not to scare [the animals] from their country." As Frenchmen noted, the requirements of intergroup harmony were simple: to arrange beforehand a cooperative and agreed-upon schedule. Meanwhile, and perhaps obviously, the "Laws and Customs" of prairie hunters prohibited one group from doing anything to scare away a bison herd that another was stalking. Such an act would have been considered, according to Hennepin, "a great injury to *the Publick*" (emphasis added), and cause for war.[65]

Surely the more important second level on which the pedestrian bison hunters' communitarian ethic operated was within the group. For this mode of bison hunting to succeed, there simply was no room for individualism or selfishness. In Liette's hunting account, for instance, he marveled about the way in which specific rules prohibited any hunter from chasing bison on his own, even in cases when he had a sure chance to make a kill. Noting that certain members of a hunting party were appointed as "guards" to prevent individuals from hunting on their own or doing anything to frighten away the animals, he witnessed one couple punished for carelessness in this respect. Part of the prehunt ritual among the prairie Indians that Perrot visited involved a chief hunter warning each member of the party to put the group interest above the member's own, with reminders of dire consequences for violations.[66]

It is no surprise that this created solidarities that did not exist among those whose subsistence rested on different resources. What is clear about this solidarity, however, is that it came with an underlying assumption of equality. No one was exempted from the imperatives to act for the community's benefit and sublimate self-interest. Even high-status individuals were

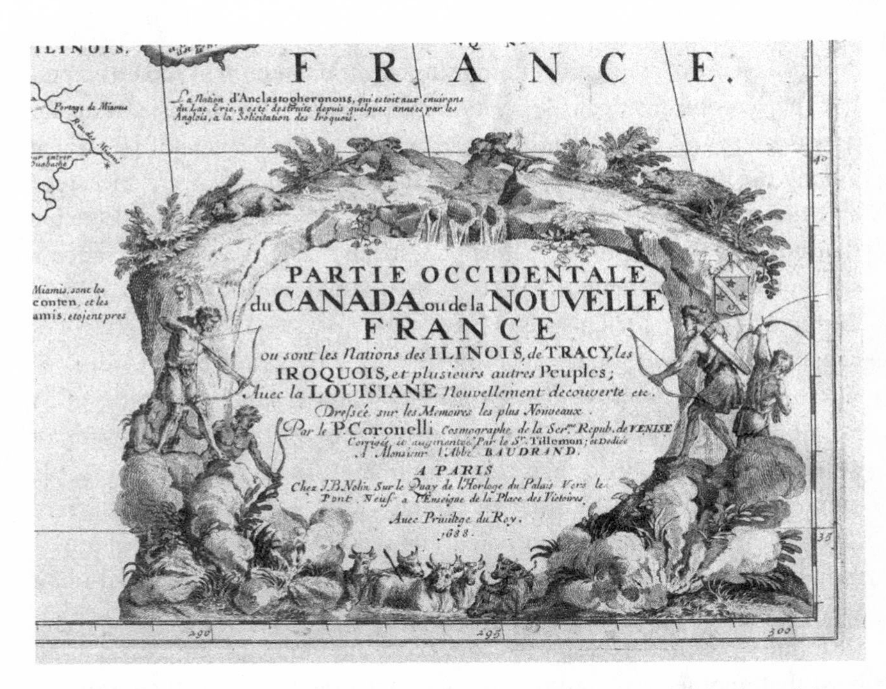

Techniques of the hunt. The lower portion of this 1688 map cartouche is one of the early European illustrations of bison hunting in the prairies. Note how the hunters appear to burn the grass to pressure the bison into a creek before shooting them with bow and arrow. Map detail from Vincenzo Coronelli, *Partie Occidentale du Canada* (Paris, 1688). Image courtesy Rare Book & Manuscript Library, University of Illinois at Urbana-Champaign.

subject to the laws about hunting. Meanwhile, the same sense of equality was embodied—literally—in the way the rewards of bison hunting were distributed. As Hennepin noted, after the hunt was over they shared the yield equitably, for "They divide these Beasts according to the number of each Family." Well-established rules governed how the meat was portioned out according to family size, and the process of distribution was conducted with "great equity and justice." And while individual hunters importantly were allowed to keep and distribute the specific animals that they individually had killed, and while "some have more and others less," nobody went hungry.[67] Finally, even in cases when one hunter amassed a large amount of meat, it is clear that the social imperatives of many prairie bison-hunting societies required him to distribute and share it. A good example comes from the episode in Liette's hunt described earlier, when the Illinois shared the result of Liette's first kill with the hundreds of hunters present in the party. As Liette learned, the sense of fellowship and reciprocity that feasts like this embodied was paramount. While on the hunt Liette was invited to as many as ten feasts per day, as different ethnic groups—Kickapoos, Shawnees, and others present in the large mixed village—feasted one another as gestures of diplomacy. He could not refuse to attend and eat lest the hosting family be "grieved."[68] Independence simply was not a core value of this society; interdependence and cooperation were the guiding principles.

Yet bison-hunting communities did indeed create social distinctions. Perhaps most important, bison hunting involved strict divisions of labor, which were most explicitly drawn along lines of gender. While men worked to stalk, run down, and slay the animals, women primarily managed the meat processing. This is an important fact, and many eyewitnesses noted how sharply it defined separate experiences within the otherwise egalitarian society.[69] Surely European eyewitnesses likely misunderstood this division of labor and mapped it onto a common prejudicial trope among mostly *male* European observers: the female Indian "drudge."[70] As many sources make clear, bison hunting seemed to conform neatly to French perceptions that Indigenous men did little work while all Indigenous women worked to exhaustion. In fact, the reality in most of these pedestrian bison hunting societies was more complicated than the Frenchmen understood and involved an ideal of gender complementarity that contrasted sharply with European notions of hierarchy and dependency.[71] Moreover, the drudge work that supposedly characterized *all* prairie women's lives was likely less a signal of straightforwardly oppressive gender order within villages as much as an actual system of out-group slavery that came to characterize prairie societies in specific periods, and about which we will learn much more later. In any event, the ideal was surely complementarity and balance. Far from the "gentlemen" lazily sitting around while their wives did all the work, as Europeans imagined, men in prairie hunting parties may well have spent much time resting, preparing to run dozens of miles in a day or recovering from the arduous work of hunting.[72] Women controlled other aspects of the bison economy—separate but equally valued tasks. Still, the sharp division of labor that characterized the way pedestrian bison hunters worked was a complex reality, at once both a part of the ethic of solidarity and collectivism and in tension with the egalitarianism described previously. This would become important to the history of these ecotone groups.

Gender was not the only way in which bison hunting—despite its communitarian ethic—created social distinctions and implicit and explicit hierarchies that cut against full egalitarianism. The hunting itself seems to have been a thoroughly and excitingly competitive arena. Accomplished hunters were respected among the prairie people, and some received special distinction and prestige for their achievements. Second only to war in terms of the honor it could bring young men, hunting was an arena of striving. It would seem reasonable that many aspects of young men's lives be oriented fully around acquiring special skill and ability as pedestrian hunters. For the Illinois speakers, for instance, observers frequently noted how boys began to "exercise themselves from their earliest youth" so as to become valuable members of the bison-hunting team. Members of these cultures staged coming-of-age ceremonies built around the bison hunt, and while care was taken to prevent young boys from becoming overly competitive, developing into good hunters was definitely an important part of the life course and a central component of both collective and personal identity.[73]

Considering all these ways in which bison hunting structured and rationalized certain social practices, then, bison hunting also helped create a worldview and spirituality in which the animals and the practices of hunting were central themes. Observing the Illinois speakers in 1712, Gabriel Marest said of their religion that "all their knowledge is limited to animals."[74] It is clear that the prairie peoples' relationality with animals was expressed and even embodied through practices that were specific to the skills and techniques of pedestrian bison hunting. Especially important was the skill that so many European observers noted as paramount in these cultures: running. Running was at the center of life. Prairie people had contests over running; footraces in particular seem to have been an especially important aspect of community recreation among several prairie groups.[75] In ritual life, running was definitely a central theme, as prairie groups held ceremonies to make hunters run faster, entreating their bird manitous to assist humans to "go the same speed in running as you do in flying," according to Liette.[76] Prairie people told stories about great runners, as in the Ioway Indians' legends of Dore and Wahre'dua, two runners who challenged supernatural beings called *Hompathrótci* to footraces and won, or who ran around the world.[77] Perhaps most conspicuously, the Meskwakis and Kickapoos, along with their Dhegiha neighbors like the Osages and the Kansas, developed a vast ritual cult around the *A'ckâpäwa*, or "ceremonial runners," a guild of runners of high status and spiritual power who performed special feats of negotiation with the spirit world and with animals.[78] Running was even entangled with spiritual quests and dreams, and running itself was an act by which people could traverse both this world and other spirit worlds. It is clear that many prairie peoples' very identities—both collectively and individually—were wrapped up with the particular act of running after bison, with the activity existing somewhere between pragmatic necessity and communal ritual.[79]

Readers who are athletes may appreciate aspects of this running culture best of all, knowing the ways physiology, psychology, and imagination can come together in demanding feats of sport and movement. Given the style of distance running that bison hunting demanded, it is easy to imagine how running may well have produced distinctive physiological and psychological effects—a transcendental state at the boundary of exhaustion and euphoria.[80] But if it is important to see the skills of bison hunting as perhaps more than just instrumental physical work, and as something closer to worldview-shaping ways of being, as the anthropologist Tim Ingold might suggest, we should think of them as *more than human* as well.[81] After all, it would be hard to imagine many aspects of these prairie peoples' culture—and particularly running—existing as they did without the animals. Recall that the prairie peoples stood out to Europeans for their running prowess, with nearly every eyewitness who encountered them remarking on this conspicuous aspect of their culture. It was an undeniable fact, as Liette wrote, since watching the Illinois hunt on foot could "not but give pleasure to a thousand people themselves trained runners."[82] As Charlevoix put it, "the Illinois are perhaps the swiftest footed people *in the world*; and there are none but the Missouris who can dispute this piece of excellence with them" (emphasis added).[83] According to Hennepin, who watched some Mississippi Valley hunters chasing bison to exhaustion, "there is not a buffalo that they cannot run down," and many could even reportedly run a deer to exhaustion.[84] The talent was shared by men and women alike, for even while carrying a heavy burden, Illinois women could reportedly "run as swiftly as any . . . [French] soldiers with their arms."[85] To de Tonti they were "quick, agile, and brave," and to Claude-Charles de La Potherie they were "excellent pedestrians."[86] Antoine de la Mothe Cadillac, in writing of the Myaamias in 1718, stated that they were "real and true greyhounds."[87] All these physical acts were more than just impressive feats. To the prairie people themselves they were indeed acts of community, acts of human labor constituting important solidarities among people and between people and the nonhuman creatures on whom they centered their lives. All this went beyond subsistence and practical necessity, and indeed went beyond the human. Bison helped constitute prairie peoples' communities. They *made* prairie people.

◆ ◆ ◆

We do not really know the whole story of how the prairie people fully became bison people, nor precisely when it happened, since any modern analysis relies on the method of upstreaming. It nevertheless seems reasonable to think of this process of *becoming* as a crucial historical event that took place gradually, probably sometime in the early 1600s. The timing is important. Historians of early America have often identified the period around the beginning of European colonization as a time when new peoples were created and new identities were born, a process sometimes referred to as ethnogenesis. Frequently historians of Indigenous America center their stories of ethnogenesis on colonial contact, as in the case of the Choctaws or the Catawbas—peoples whose identity was created in part through contact and the processes of consolidation and differentiation that happened in the course of colonial encounter.[88] Recently, however, historians have explored cases in which ongoing Indigenous trajectories and environmental changes led to the creation of new peoples, or these processes of ethnogenesis, in their own right.[89] Historians have long acknowledged that something distinctive was happening in the prairie region—one scholar calls it a "bison revolution"—for the new peoples beginning to occupy the Illinois Valley and surrounding ecotone region around 1600.[90] We probably can understand that these changes and this new lifeway amounted to a process of ethnogenesis—or perhaps multiple ethnogeneses—as well.

We know that in the wake of Cahokia's decline the Illinois Valley was occupied by small groups of people organized into modestly sized villages. Having abandoned, avoided, or altogether replaced the large settlement centers of the Mississippians, the inhabitants of the eastern prairies in the 1600s were proto-Siouan Oneota peoples as well as proto-Algonquian migrants who began to arrive from the east around 1450. We know that these Oneota and proto-Algonquian peoples "moved down" the trophic levels in their subsistence, abandoning—if they had ever committed to it in the first place—the narrowly focused system of intensive corn agriculture for a wider base of resources, including a rich cycle of seasonal hunting and gathering. In terms of political structure, small tribe- and village-based social organizations were the hallmark of these societies. It is impossible to say for certain how they thought of themselves or whether they prioritized coherent tribal, village, or possibly patrilineal clan identities like the ones that came to characterize the prairie people of the contact period. Overall, the transition from the Mississippian period featured a process of "factionalization into smaller economic units," as newcomer groups and ancient inhabitants adapted to the new environment and new opportunities.[91]

But then came the bison revolution, and it changed everything. New energy made new people.[92] Of course this was different from plains groups who adopted horses and for whom such a choice meant abandoning the stability of their previous lifeway. Those people tied themselves tightly to a single resource that eventually constrained options and led to dependency. For the pedestrian bison hunters of the ecotone, however, the safety nets of their various subsistence strategies remained intact even after they began pursuing the animal. It's an important reality that suggests a larger underlying truth: the shift to bison hunting in the ecotone was about choice and opportunity, not necessity or constraint.

Still, making this choice came with commitments and requirements. No less than the equestrians, pedestrian bison hunters made history-defining trade-offs as they sought this opportunity. Bison hunting on foot required massive work, and that work had to be organized. It fostered certain behaviors and it fostered certain worldviews. It created special practices. All of this was new. Arid and straightened circumstances of the Pacific Drought that challenged Cahokia beginning in the 1200s probably had helped shape the small-scale and loosely organized groups that rose in its wake. With the new abundance of the ecotone in the Little Ice Age—epitomized particularly in the herds of bison—the equation shifted. Our intuition is simple: bison became constituents of the prairie peoples' newly complex social worlds and *made* new societies.

This history was not a case of straightforward environmental determinism but nor was this a simple matter of human mastery. Instead it was something more complicated: the co-constitution of human and nonhuman lives as they tangled together. For the people of thetallgrass ecotone, moreover, these consequential changes happened right in the moment before the arrival of colonists, which is to say in the decades just preceding when standard tellings of their history usually begin. Almost every group that the French encountered upon their early ventures into the ecotone region, and particularly the Illinois and the Meskwakis, had recently entered the tallgrass prairies and become pedestrian bison hunters. Oneota-descended Siouan peoples had been present in the region longer, and had similarly remade themselves around a bison lifeway in the eastern prairies. Over the course of just a couple generations, and on the heels of an important species shift, the bison had become perhaps the most important agent in shaping how the history of these Indigenous groups unfolded.

This dynamic would not stop with the arrival of Europeans. Having thoroughly shaped what it meant to be Illinois or Meskwaki in the precontact history of the ecotone, bison and other nonhuman entities were about to shape what the opening moments of colonialism looked like in this edgy region. Entangled together, co-constituted together, the prairie people and their animal partners of the ecotone were *running* forward, into history.

4

EDGE AND WEDGE

IN THE FALL OF 1680 A WAR PARTY OF SOME 800 IROQUOIS TRAVELED hundreds of miles from western New York to the eastern prairie region to launch an attack against the Illinois speakers, one of the great turning points in the Indigenous history of the Midwest. The target was no small camp, but rather a consolidated population center where various bands of the Illinois had been living together for most of a decade. The Iroquois had timed their expedition carefully, and they spent October patiently following the Illinois, who slowly retreated down the Illinois Valley to the Mississippi River, where they separated into smaller villages. Now the Iroquois advanced. The battle was lopsided: Illinois warriors from the Tapouaro band reportedly ran away and the result was a rout. Outnumbered, some 350 Illinois were killed, and somewhere between 350 and 700 women and children were led away as captives. Regrouping, the Illinois counterattacked, and the violence rippled over years. It was one of the most consequential moments of military violence in the seventeenth-century ecotone region.[1]

In explaining these events many historians have emphasized a simple context: the Iroquois' contact-era *invasion* of the western Great Lakes, begun in the mid-seventeenth century and still ongoing in 1680. It was the Iroquois who, triggered by losses from European disease and motivated by a quest for hegemony in the ongoing Beaver Wars of the seventeenth century, over many decades assembled unprecedented military expeditions like this one to seek captives among Indigenous people of the western Algonquian world, marching their victims back to their villages to be tortured or adopted as replacement kinsmen. Living in a region far distant from the main action of colonialism and following a lifeway that had little to do with beaver hunting, the Illinois hardly shaped why this violence happened at all. The 1680s were a time when the Illinois and their prairie neighbors found themselves caught up in forces originating from far away, an "engine of destruction" spreading across the continent.[2] Their role was to respond and react, to flee and collect in refugee centers, and ultimately to decline. Like so much of midwestern Indigenous history, the usual narrative of 1680 emphasizes outsiders and invaders, not the prairie people themselves.[3]

This accepted story line is oversimplified. To understand why, return to the Illinois village a short time before the Iroquois arrived. In October 1680 the Illinois speakers assembled their own military force, nearly as large as the Iroquois party that would arrive in Illinois a few weeks later. And just like the Iroquois party, this group set out on a massive expedition, to raid "very far from their country." Led by a chief called Paessa, this party headed west on an outing that was in some ways the mirror image of the Iroquois raid. It, too, was successful, scoring a major victory on an unnamed village likely located somewhere west of the Mississippi River. Its success was only partial, however, considering that the expedition, which contained "more than half of [the Illinois'] warriors," was gone for months. Here was a fateful turn. Paessa's expedition had left the Illinois village undefended. And, importantly, sources agree that the Illinois villagers were victimized so badly in 1680 by the Iroquois precisely because they were undefended.[4]

These facts complicate our understanding of the Iroquois raid as a simple story of invasion. Indeed, it is not too great a stretch to say that the transformative impact of the Iroquois 1680 raid was caused just as much by the Illinois' own actions as by the actions of the Iroquois. And while this does not change the pathos of the situation, it does challenge our understanding of what was primarily driving Illinois history in the early contact era. For while it is clear that Paessa's raid was not wholly unrelated to the Beaver Wars or affairs in the east, it also had roots in a different history. Indeed, Paessa's raid was in many ways a local phenomenon growing out of local trajectories.

Keeping the backstory of the previous chapters in mind, we are now able to tell this new story. Beginning in the early seventeenth century, newcomers and ancient inhabitants came together in close proximity in the ecotone region and their populations grew in response to new opportunities. Formerly mostly useless for human subsistence, the tallgrass prairies of the Midwest in the 1600s became a boon, thanks to climate changes and bison. Adapting themselves to the new economic possibilities of pedestrian bison hunting, diverse groups of people made prosperous lives and invented whole new community dynamics as they exploited suddenly available energy streams embodied in "innumerable" herds of animals. Hardly a static or stable Indigenous world just waiting for history to begin, here was a region where dynamic change and considerable innovation had been the rule for generations. Well before the violence of the Beaver Wars and colonial epidemics sent refugees to the west, and well before Iroquois invasions like those in 1680 visited colonial violence directly on midwestern populations, the opportunities of bison hunting were already helping shape dynamic change and instability.

The groups inhabiting the ecotone in the 1600s were a diverse mix of Siouan speakers and Algonquians, both newcomers and ancient residents. Viewed at the level of individual villages, these groups had intact and coherent traditions, even if ones that were obviously evolving in response to recent circumstances. Viewed at a larger regional scale, however, and considering relationships between villages, the people practicing these new lifeways and chasing the new opportunities of bison hunting in the tallgrass prairies had considerably less coherence. For while they needed to get along, given their mutual occupation of this country and the cooperative demands of bison hunting, these were people who in many ways had little in common—no longstanding diplomacy, no patterns of stable intergroup relations. Pulled together by new opportunity, they existed side by side within a landscape of strangers.

In this context, competition and conflict were perhaps inevitable. They manifested themselves in a fairly typical pattern of what the French knew as mourning war, especially as the first effects of colonialism did start to arrive at least as early as the 1640s. But while this kind of captive-oriented warfare was common to most Indigenous groups throughout the seventeenth century Great Lakes region, in the ecotone it took on a special cast and a special function. Given the ecotone residents' need for new intergroup relations, and given the ready availability of "strangers" in their borderland regions, prairie Indians like the Illinois now made captives the center not only of war but also of a special kind of peacemaking: captive exchange. Exploiting the special social diversity of their region, groups like the Illinois and the Sioux sought captives *among* strangers in order to gift captives *to* strangers, using the symbolism of the subordinated captive to define the solidarities and boundaries of their tentative social networks. The practice was not unique, but it was particular. Unrestrained by long-standing alliances and seeking ways to separate friends from enemies in their complex borderland, they used the practice of mourning war and captive exchange on a large scale to shape a volatile regional order featuring uneasy and shifting coalitions. In so doing they simultaneously exploited the key tensions of intercultural life in this newly diverse region and also looked to resolve them, relying on captives taken from certain neighbors as the "glue" to hold together new alliances with other neighbors.[5] Prairie people became specialists in captive-raiding and slave-trading.

This important function of captive-raiding in the ecotone was almost certainly the crucial logic behind Paessa's raid in 1680, the key explanation for why the large group of Illinois left their villages undefended in order to raid "very far from their country" at such a fateful moment. Importantly, when the Illinois returned from their raid to find their own decimated villages, they had with them a large number of captives taken during their own expedition. Other evidence suggests what they probably intended to do with them: send them on to allies to signal and seal alliances. And if that is true, this special history of captive-raiding in the ecotone is every bit as central to our understanding of what was happening in the Illinois Valley in 1680 as the Beaver Wars. Far from simple victims of invasion, residents of the ecotone such as the Illinois were shaping their own patterns of warfare, exploiting special imperatives and opportunities rooted in recent ecotone history. By imagining an "engine" powering history from the east, the typical story of colonial midwestern history ignores key trajectories happening in the Indigenous west itself, trajectories which constituted if not exactly a separate engine then certainly a distinctive and specific transmission, affecting how phenomena like mourning war and disease drove history. To understand how and why the Beaver Wars rippled across the continent to the remote Midwest we need to understand the special circumstances of the bison people. The story has never been told in its full complexity.

◆ ◆ ◆

To understand how the beginnings of colonialism affected Indigenous history across the Great Lakes in the contact era it is necessary to start with the crucial logic of mourning war. Within many Indigenous cultures, when any family suffered the death of one of its members, a deeply felt imperative motivated survivors to try to compensate for the loss of that family member in battle. This imperative to satisfy missing kinsmen was a duty—a singular and driving force in Indigenous military culture. Once factors like exogenous disease and early colonial violence caused population losses among eastern Indigenous groups at first contact, this logic motivated such devastating warfare among Indigenous groups, as battles led to more battles and warfare to further warfare.[6]

The logic of mourning war speaks to the central preoccupation of Indigenous life: kinship. Although we often think about Indigenous history in terms of collectives like "tribes" and "confederacies," it is clear that many Great Lakes people around the contact period understood their social lives much more through categories like clans and—especially—extended families. The kinship construct organized intimate lives among close nuclear families, and it also organized more distant relationships among people who were not biologically related but whose relations were conceptualized in terms of large and extended—and often fictive—kinship networks. When a family member died, the loss was felt not just at the level of the household. It was understood to extend much more broadly, threatening the connections and harmony among large groups of people. The logic of mourning war was to restore that balance and those connections by ceremonially "raising the dead."[7]

Like European warfare of the early modern period, mourning war was often focused on taking casualties or inflicting death on enemies. Given the emphasis on replacing kin, however, a more satisfying way for a family to compensate for a deceased relative was not by inflicting death on an enemy but rather by controlling an enemy's life. These people—captives—could then be incorporated into the village to "replace" the dead. As Liette estimated, "the man who brings a prisoner to the village is more esteemed than the one who kills six men among the enemy." Father Sébastien Râle agreed, writing of the Illinois that taking captives was the most important goal of warfare: "When [an Indian] returns to his own country laden with many scalps, he is received with great honor; but he is at the height of his glory when he takes prisoners and brings them home alive." The priority on captive-taking for Indigenous groups of the Great Lakes and prairies is reflected best in the many ceremonies and dances in which men boasted about their captive-taking successes, telling stories about the number of captives that they had taken and how it was done.[8]

The complicated fate of the captives was for Europeans perhaps one of the most incomprehensible aspects of mourning war. Upon returning to the village, the war party would consult together and decide "what to do with the prisoner." Often the prisoner would be tortured and then ritually executed. Eyewitnesses to this feature of mourning war usually did not understand its ritual significance or the catharsis and transcendental release experienced by the villagers as they relished the enemy's powerlessness. Among the Illinois, as Sebastien Râle noted, "This reception is very cruel . . . Some tear out the prisoners' nails, others cut off their fingers or ears; still others load them with blows from clubs." It was during this initial moment of arrival to the village that captives frequently were disfigured with "marks"—nose cropping, brandings, and other painful humiliations aimed at subordinating the victim and demonstrating the prowess of the captors. If captives were chosen to be killed, now they were painted and prepared for the flames.[9] Execution of the captive was the ultimate act of superiority for the conquering village.

Considering their function, to "resuscitate the dead," however, many captives were not executed. Rather, they were incorporated into the village, converted from outsiders to insiders, then accounted as part of the group. This was a complex process, with many variations. In most cases captives "granted life" were brought in and made to assume a degraded position in the village. Even if their function in part was to "replace" kinsmen, captives were frequently understood as something closer to property than true kin.[10] The close equivalence between words for "prisoner" and "slave" in many Jesuit translations of Great Lakes Indigenous languages suggest the gist of captives' status. It was never totally equivalent to the notion of chattel slavery, but the position assumed by many captives in their new kinship lineages certainly implied subordination and less than full membership. Indeed, a common metaphorical connotation in Algonquian languages liken prisoners to domesticated animals or pets—like dogs. It is clear that many captives were indeed made to work for the family into which they were incorporated; for many captives, their fate in their new family was to be treated more or less "as a slave."[11]

But not always. Some captives were brought into their new community on a less-degraded footing, and then "adopted." The ways and means of captive adoption were complex, and hard to generalize, although it seems clear that prisoners sometimes were made to assume new identities and become the kinsmen they were replacing. This transformation was symbolized by rituals performed at the welcoming of the captive: stripping, washing, reclothing, painting, and especially renaming, as captives were given the name of the deceased person they were meant to replace.[12] To be sure, such identity transformation seems to have been much more likely—and feasible—in cases when families looked to existing allies or friends, rather than prisoners, to "take the place" of deceased relatives. Nevertheless, it is clear that some prisoners were adopted and trusted to occupy regular and even high-status positions in families, becoming "the same as he whom he replaces and whose name he carries." And if this was unusual, there were other ways prisoners entered families in positions that approached full kinship. In the case of women, for instance, captives seem to have been incorporated into polygamous families as second and third wives, a role that gave them full identity as family members but within a certain hierarchy.[13] Captives adopted as children seem to have the most chance of becoming full members of the family. In general it seems that some captives could "earn" a higher status and over time rise out of an initial degraded position of slavery.[14]

Given the complex meanings and ambiguous transformations it entailed, it is not surprising that powerful rituals and ceremonies surrounded the ways and means of mourning war, from the ceremonial initiation of a captive-raiding expedition to the complex treatment of captives upon their entry into the village. And while French eyewitnesses often foregrounded violence in many accounts of this ceremonialism, it is clear that the major theme was also community. This is important. At the beginning of the mourning war, families hosted an extensive feast, recruiting members of the village and enticing them—"enlisting" them—in the collective violence. Upon bringing the captive home, the whole village participated in the ritual of incorporation and the violence directed at the victim was thoroughly cathartic for each member of the village, not just members of the specific household welcoming the captive.[15] Much evidence suggests that families coordinated to help perform the ritual of incorporation, painting bodies, stripping the captive, and laying out the elaborately painted robes with which prisoners were cloaked.[16] As one Illinois elder put it upon giving a feast on the departure of an expedition, all of this community participation was not incidental. A missing kinsman represented a loss to one and all, and his replacement was a collective process. "He was your relative as well as mine, since we are all comrades," as one Illinois man expressed it.[17] As the logic here suggests, it was a community imperative to restore missing kinsmen.

An important but subtle part of this community dynamic in mourning war was the transaction between those who received the new captive and those who actually captured him or her. Not everybody could be a great soldier and not everybody could go to war. And this fact structured another important meaning of mourning war, as the captive became not just replacement kin to a family but a gift from one person to another. We see this logic at work at the start of an expedition, when families in mourning feasted soldiers to entreat their help, and with soldiers in turn declaring themselves "ready to die" to restore their neighbors' kinsmen. Returning to the village, these soldiers subordinated whatever self-interest they might have regarding the fate of their prisoners, turning their captives over to elders to "decide what they will do with the prisoners and to whom they will be given." Next there would be elaborate consultation with eligible families, those who had "relatives [who] have been killed by warriors of the nation from which they bring back prisoners." Having made the decision, the committee would approach each family with the happy news: that "they are delighted that the young men have brought back some men to replace, if they desire it, those whom the fate of war has taken away." Upon learning of their good fortune, the receiving family now would express its debt to the soldiers: "For this offer great thanks are returned."[18] Here was one of the most important moments in the whole process. To be sure, mourning war's primary function may well have been to restore balance and mend relationships *within* family lineages, relationships between survivors and lost kinsmen. But another crucial logic of mourning war was to strengthen relationships *among* lineages, relationships expressed by the act of capturing, gifting, receiving, and incorporating prisoners. As we will see, these meanings of the gifted captive would become especially important to the history of the ecotone.

Whether subordinated as slaves or adopted as true kinsmen, captives theoretically could restore balance in kinship groups and enlarge families. At the same time, it is easy to see how the imperatives and logics of mourning war interacted with colonialism to produce devastation, or what Richard White calls an "engine of destruction." As soon as exogenous diseases began to hit Indigenous villages in the contact era, the almost mechanistic cycle of violence started spinning. Trying to replace kinsmen lost to epidemics, Iroquois and other eastern groups mounted large raids on their neighbors. Victims collected in large villages for defense, where disease epidemics proved only more deadly. To make matters worse, prisoners taken from one village to another likely became disease vectors spreading more disease. Guns only worsened the violence. Adding to the toxic cycle and producing nefarious feedback loops, disease epidemics and raids compromised villagers' ability to practice normal subsistence and maintain basic health standards, adding to severity of disease. Entangling in a vicious cycle, many of these factors became simultaneous causes and effects of one another, provoking even more casualties, more missing kinsmen, more mourning war.[19]

The history of how mourning war began and spread particularly at the hands of the Iroquois is well known. In the first half of the seventeenth century the Iroquois raided their neighbors to the west, focusing especially on the country of the Huron, east of modern-day Lake Huron, as well as along the St. Lawrence Valley. In the mid-seventeenth century the results of those expeditions were obvious, as eyewitnesses noted more captives than native-born villagers among the Iroquois. Soon this created an obvious problem. Holding so many captives—well over half of a village's total population—was problematic. As one Iroquois ethnologist puts it, when assimilating captives as new villagers, the Iroquois used a strategy known as *we-hait-wat-sha*, which translates to "a body cut into parts and scattered around."[20] The idea refers to the effort to weaken and eventually erase captives' previous identities and replace them with new relationships and new allegiances. But this was a fraught practice, not just because the captive preserved social ties and the desire for home, but because the captives' original kinsmen remembered and longed to repatriate their lost kinsmen as well. These were people, like the ethnic Hurons, "whose mothers had escaped from the ruin of their tribe when the Irroquois had invaded their former country."[21] Importantly, many survivors did not forget these ties. Meanwhile, given the way diplomacy worked through practices of adoption and fictive kinship among Great Lakes Indians in the 1600s, most neighboring villages were usually full of relatives—people already connected by kinship and fictive kinship ties. They could not be slaves.

The difficulties involved in holding captives from neighboring villages, together with the need for more numbers, caused the Iroquois—and then, in turn, their victims—to press further west after their initial expeditions in the 1640s. In the far western Great Lakes lived people with whom the Iroquois did not share kin and whose location far from the Iroquois homeland meant that future entanglements between assimilated captives and their surviving kinsmen and allies could be kept to a minimum. This was especially true as Iroquois war parties took their expeditions all the way to modern-day northern Wisconsin, where they raided on villages of traditional enemies like the Ottawas, but also the Meskwakis, the Illinois, and many other groups ostensibly far removed from the politics of the Beaver Wars. For Nicholas Perrot, these long distance raids merely reflected the Iroquois' essential qualities as "pitiless man-eaters, [who] have always taken pleasure in drinking the blood and eating the flesh of all the different tribes" and who thus "seek their prey even to the confines of America."[22] But one benefit of raiding so far from home was that they were doing so in a country of no kin—of strangers. Bringing home such captives allowed the Iroquois more easily to practice we-hait-wat-sha.

As the Iroquois pushed west, however, they entered a region defined by diversity and recent change. On the one hand, some of the people already there were actually refugees from the east. Reaching Green Bay in 1653, for instance, the Iroquois encountered and attacked a village comprised primarily of Hurons and Ottawas, recent arrivals from the east, themselves driven there by the early phases of the Beaver Wars. The further west the Iroquois went, however, the more they encountered people whose presence in the prairie region was part of a different history rooted in recent environmental changes and migrations. As the Iroquois attacked such groups as the Illinois in 1655, the Illinois retaliated and their history soon became part of the story of the Iroquois wars, as historians have shown. But in fact, these Iroquois attacks only added new cycles of mourning war to an ongoing set of conflicts. That is because for the westerners themselves, no less for the Iroquois invaders, the ecotone region was full of strangers, of non-kin.[23]

◆ ◆ ◆

To begin to understand on its own terms the violence that broke out in the prairie ecotone region in the contact era, we have to start by observing a basic fact: the people of the ecotone were not just strangers to the Iroquois but also in many cases to each other. The Ho-Chunks—Chiwere Siouans and descendants of Oneota hunters—occupied present-day eastern and central Wisconsin and had likely the longest tenure in the region. Sometime in the early seventeenth century the Algonquian-speaking Meskwakis arrived, moving from the area around modern-day Detroit. Their migration may have been inspired by what historians have called the "persecution" of Iroquois attacks at the beginning of the Beaver Wars, but evidence from oral tradition and other sources suggests that they had departed modern-day Michigan well before those conflicts began.[24] The Miami-Illinois speakers—likely the bearers of the Danner Series of pottery that reveals their migration from the Ohio

Valley—began their in-migration into the Illinois Valley and points west sometime in the early 1600s; they either displaced or incorporated with Oneota groups who had dominated the Illinois Valley since Cahokia. On the western side were groups like the Mdewakantons, the Sissetons, the Wahpekutes, and the Wahpetons—collectively known as Dakotas—as well as the Lakotas, known to the French as "Nadouessi." These Siouan peoples had moved from the central Mississippi Valley to the eastern prairie ecotone of modern Minnesota over the long course of the Mississippian and post-Cahokia period.[25] Living near modern-day Chicago were the Potawatomis, who established several villages, also relatively recently, during a diaspora from the northeastern Great Lakes region. These were the largest groups in the region, which were generally organized in loose village- and clan-based polities; other groups included the Mascoutens, the Menominees, the Sacs, the Kickapoos, and many others besides.[26]

French people knew little of these peoples' history when they first traveled west in the second half of the 1600s. What they did know, as Allouez wrote of the Meskwakis, in a sentiment echoed repeatedly by French eyewitnesses of these groups, is that they were "settled in an excellent country."[27] In some respects "settled" was surely the correct word, for these groups—including the long-resident Ho-Chunks and the Sioux, as well as the relative newcomers like the Meskwakis and others—were fortunately rooted to their special geography. One early French account, for instance, estimated that the Illinois numbered one hundred thousand, living in sixty villages in and around the Illinois Valley and points west, and supported an army of twenty thousand men. This was an exaggeration, but the true population of Illinois-speakers was still quite large, probably around twenty thousand. The Potawatomis, having just moved to Green Bay, were a significant portion of the Anishinaabe Three Council Fires, thousands of whom moved west from the upper Great Lakes to the prairie edge, according to Jesuit reports. The Ho-Chunks were legendary for their might, and eyewitnesses estimated their population at between four and five thousand men, which would suggest an overall population of 20,000, while the bulk of the eastern Sioux comprised at least twenty-four thousand. In the 1660s the Meskwakis numbered somewhere around five thousand people, and the Miamis were estimated at four to five thousand men, or likely well over ten thousand in total.[28] In short, these were "populous nations" living in the ecotone, "a very attractive place, where beautiful Plains and Fields meet the eye as far as one can see." It was a distinctly different environment and still largely remote from the effects of colonialism in the east; French Jesuits considered the Sioux's country to be "well-nigh at the end of the earth, so they say."[29]

Although probably intended half-jokingly as a reference to the difficulty of missionizing among the remote Sioux, there is nevertheless something apt about the Jesuits' notion that the prairie people lived at the end of the earth, or virtually in another world. For, just like the ecology of their region, the social patterns of their region were also new, unsettled, and in flux. This is best reflected in the story that French diplomat Nicholas Perrot learned about the western groups and their precontact history as he traveled in the region in the mid-seventeenth century. According to Perrot, each of the western groups had long previously "inhabited a region that belonged to him; and there they lived with their wives, and gradually multiplied." As their populations grew in the distant past, however, "they separated from one another, in order to live in greater comfort." Indeed, migration helped create prosperity, as they moved into the sparsely settled and abundant ecotone environment. But there was a problem: "They became, in consequence of this expansion, neighbors to peoples who were unknown to them, and whose language they did not understand." Perrot's story here points to the important linguistic barrier between mutually incomprehensible Algonquian and Siouan language families, a divide that was as important to the westerners as the one between Iroquoians and Algonquians in the Great Lakes. Finding themselves next to unintelligible strangers, some of the newcomers developed animosity: "Some of them continued to live in peace, but the others began to wage war." The wars inspired more migration, as certain groups "retreated to more distant places, where they found tribes whom they must again resist." Here was the crucial background history of the western Great Lakes and eastern prairies: the social consequence of arriving and persisting in these *more distant places*. And despite how the bison-hunting economy created imperatives for intergroup community, rivalries and tensions certainly simmered among the newcomers and the long-resident groups.[30]

Importantly, and as historians have begun to recognize, the tension that defined intergroup relations in the prairie ecotone was a consequence of this recent history and the social flux produced by reshuffling and resettlement in the wake of Cahokia. In much of the Great Lakes, intergroup harmony rested on individuals from different villages intermarrying, becoming kin, and adopting one another to unite clans and mostly patrilineal kinship groups.[31] By becoming relatives in these ways, many groups could remain distinct but also harmonious with their neighbors. But the recent history of the ecotone meant that this process was only just beginning among the prairie groups. Moving into these new zones, people like the Illinois generally did not share kin with new neighbors like the Ho-Chunks or the Meskwakis. Far from merging kinship and clan lines, many of the main groups—the Sioux, the Meskwakis, the Ho-Chunks, the Illinois, and the Mascoutens—remained largely unrelated. When rivalries took shape between and within villages, kinship provided no "glue" to hold this world together.

While colonial factors like the Iroquois wars would add to it, this social flux was not just a matter of "Iroquois-induced chaos," since it predated the Iroquois incursions of the mid-seventeenth century.[32] At root was the resentment felt by older residents like the Ho-Chunks, as newcomers arrived in the region and invaded on territory in which the Ho-Chunks considered themselves sovereign. Indeed, it makes sense that the Ho-Chunks, the Oneotas' descendants and in many ways the "founders" of the bison-based lifeway in the eastern prairies, would feel a sense of their considerable power in this special region. But, as French chronicler Claude-Charles de La Potherie understood, the Ho-Chunks' dominance did not remain uncontested; as newcomers arrived in the region, the Ho-Chunks asserted themselves. As La Potherie put it, "If any stranger came among them, he was cooked in their kettles." With the influx of migrants, this kind of violence was not unusual, and the Ho-Chunks developed a reputation among neighbors for "tyranny" and a self-image as "the most powerful in the universe."[33]

Their new neighbors did not agree. Before long a cycle of violence—a series of mourning wars, in this case powered by distinctively western historical trajectories—took off. Exposed to "vigorous" attack, the Ho-Chunks "were compelled to unite all their forces in one village," where fatefully the Ho-Chunks suffered an epidemic, perhaps the first so-called virgin soil epidemic this far west in the contact era, reducing their population by 80 percent. Reeling, yet more determined than ever to "satisfy the manes of their ancestors," the Ho-Chunks struck a group of Illinois who came to their village on a diplomatic mission in 1645, surprising them in a "massacre." And now the violence spiraled further. For their part, after taking two years to mourn their own losses from this ambush, the Illinois returned to the Ho-Chunks' village with a similar agenda, aiming to take captives to compensate for their own losses. And so they did. "On the sixth day [the Illinois] descried [the Ho-Chunk] village, to which they laid siege. So vigorous was their attack that they killed, wounded, or made prisoners all the [Ho-Chunks], except a few who escaped." Apart from captives whom the Illinois "had the generosity to spare," many survivors did not last long, for they were "severely wounded by arrows."[34]

To be sure, this was typical mourning war, as La Potherie understood. Among the Ho-Chunks and the Illinois, no less than among groups further to the east, the prairie people were motivated by "the notion that the souls of the departed, especially of those who have been slain, cannot rest in peace unless their relatives avenge their death." And yet, what La Potherie likely did not understand, and what later historians have never really explained, is how the forces creating the "engine of destruction" in the seventeenth-century western Great Lakes—the interrelated factors of captive-raiding, mourning war, and exogenous disease—stood apart from the Beaver Wars and stemmed instead from the recent history and particular local circumstances of the prairie Indians. The difference is not the powerful logic of kinship replacement but rather the complete lack of restraints to the resulting violence. Everybody involved in these attacks and counterattacks were strangers to one another. This is why the Ho-Chunks attacked not just the Ottawas who attacked them initially, but the Meskwakis and others as well, soon "declar[ing] war on all nations whom they could discover." It also is likely why the Ho-Chunks attacked the Illinois, despite the fact that the latter brought them food and tried to show them "compassion." It is why these conflicts could not be reconciled, despite the Ottawas' and later the Illinois' efforts to send "envoys" and "form a union" with the Ho-Chunks. And above all, it is why the violence reached the scale that it did, with the Illinois killing or taking captive nearly *the entire population* of the Ho-Chunk village in an attack around 1650. We can best understand these events when we consider them in the context of regional realities: the migrations shaping the ecotone region and of course the environmental changes to which they were related.[35]

Events like the wars of the 1640s between the Illinois and the Ho-Chunks were only too common in the prairie ecotone of the seventeenth century. A protagonist in much of the fighting were the Lakotas, the easternmost families of the Sioux. Perceiving themselves as similarly "invaded" by newcomers from the east, the Sioux launched a series of mid-seventeenth-century attacks against the Ottawas, the Hurons, the Illinois, and the Meskwakis. Importantly, when Ottawa and Huron explorers arrived in Sioux country in 1656, the Sioux first tried diplomacy, repatriating the captives they had taken among a group of naive trespassers. Seeking access to new French trade goods brought into the western region by the earliest fur traders, the Sioux reasoned that "that these new peoples who had come near them would share with them the commodities which they possessed." Indeed, on several occasions the Sioux went out of their way to form bonds with the newcomer groups, particularly the Ottawas. But the Ottawas mistakenly viewed the Sioux diplomacy as evidence of weakness and sensed opportunity in the ecotone, particularly the "buffaloes and other animals [which] were found there in abundance." Rejecting the Sioux diplomacy, they contemplated driving the "Sioux from their own country in order that they themselves might thus secure a greater territory in which to seek their living."[36]

This competition for territory in the rich ecotone region to the west soon created conflicts as important as any Iroquois invasions from the east. In 1670–71 the Ottawas organized an invading force of one thousand among newcomer Algonquians including the Sauks, the Meskwakis, the Kishkakons (bear clan of the Ottawas), and the Hurons, and they even built a fort in Sioux territory. Now the Sioux responded and showed their might. As Perrot told it, "The Sioux pursued [the invaders] without intermission, and slew them in great numbers, for their terror was so overwhelming that in their flight they had thrown down their weapons." The Ottawas and the Hurons were forced back to a location at Chequamegon, north and east of the Sioux territory, and never again attempted to invade the prairie region. But others—in particular the Illinois and the Meskwakis—remained in villages near Sioux hunting territory, and thus remained at war with the Sioux. From their position in the isolated prairies, beyond what Pekka Hämäläinen calls "the iron frontier," the Sioux slowly exercised a regional hegemony.[37]

Meanwhile, this local competition over territory was soon exacerbated and joined by the better-known invasions from the east, as well as more disease. The Iroquois came and attacked the Illinois first in 1655. In 1665, intending to attack the Potawatomis, the Iroquois landed on a village of Meskwakis, killing seventy and taking thirty captives. They attacked the Meskwakis a second time in 1670. Regional conflicts were also exacerbated as the French came to set up their trading centers and missions at Chequamegon and Green Bay, and later near the northernmost point of the Illinois Valley, giving the prairie Indians yet further reasons to compete against one another. Associated diseases introduced at these trading centers produced the impetus for more mourning war. When the French arrived, evidence of the disordered, tense situation in the ecotone social world was everywhere to be seen.[38] Yet in explaining their problems to the French newcomers, Indians of the Great Lakes only partly blamed the Iroquois or the Beaver Wars, and instead usually focused more on these western regional conflicts. Traveling through the country in 1671, Father Claude Allouez arrived in a palisaded village of the Miamis and the Mascoutens, where one host explained their most important military challenge as independent of eastern circumstances: "You have heard of the peoples called the Nadouessi. They have eaten me to the bone and not left a single member of my family alive. I must taste of their flesh as they have tasted of my relatives."[39]

Indeed, entanglements with the Sioux were the defining conflicts of the contact period for many ecotone residents. The Meskwakis went to war with the Sioux throughout the 1670s. Meanwhile, the Illinois appeared to Allouez to be quite "reduced" in 1666, owing to their own victimization by the Sioux. In turn, ecotone residents looked for revenge among other neighbors. The Illinois attacked Meskwaki villages in the 1670s, making "raids upon them, and Carr[ying] off others into captivity." Mesquaki peoples told the Jesuits, "take pity on us . . . protect us from our enemies."[40] But "our enemies" meant not just the Iroquois but practically everyone, especially the Sioux, and the whole region was volatile. Exaggerating the Illinois' victimhood, Allouez asserted in the 1660s that "continual wars with the Nadouessi on one side and the Iroquois on the other [have] well-nigh exterminated [the Illinois]." While the notion that the Illinois had been "exterminated" was a wild overstatement, continual wars with the Sioux—whom the French began to refer to as the "the Iroquois of *this* country"—shaped history in this period as much as any eastern invasions.[41]

The ecotone residents needed a way to resolve these animosities, and in this violent context a special logic for peacemaking did begin to emerge. Ironically, it was rooted in the special dynamic of Indigenous warfare itself. As the engine of destruction spun, the ecotone residents may not have shared much common ground or common goals of diplomacy. Yet the Ho-Chunks, the Meskwakis, the Illinois, and the Sioux had one language they did share, and that was the language of mourning war, "the notion that the souls of the departed, especially of those who have been slain, cannot rest in peace unless their relatives avenge their death."[42] The solution to the problem of intergroup violence was ironically present in the very product of that violence: captives. And given the distinct social diversity of the ecotone that grew in the wake of generations of migration, such captives were easy to come by. It was a basic matter of "availability." What set the ecotone residents apart was that they all lived in a complex borderland and, as historian Brett Rushforth puts it, "within reach of potential slaves."[43]

◆ ◆ ◆

Surely the best reflection of the search for intergroup harmony among these unrelated neighbors in the protohistoric ecotone region was the development and spread of the calumet ceremonialism that happened right at this particular moment in the eastern prairies. Everywhere French people went they witnessed ecotone residents performing this "mysterious" ritual designed to help ease intercultural relations. As archaeologists have shown, after its "invention" in the eastern Plains sometime around 1500, the calumet spread throughout the eastern prairies of the 1600s. While it soon became common throughout eastern North America, it is no accident that its origins and early spread took place in the eastern plains and especially among the groups of the prairie ecotone.[44] As Perrot put it, the Indians "of the prairies have the utmost attachment for it, and regard it as a sacred thing." Along with bison hunting and their special running acumen, the calumet was the most frequently commented characteristic of these societies noted in primary sources from the period.[45]

While diverse and multifaceted, one important feature of calumet ceremonialism was an adoption ritual meant to create fictive kinship among individuals of different villages, clans, lineages, or ethnic groups. This adoption symbolism gave the calumet ceremonialism its considerable power. After sharing the calumet, as one Jesuit put it, newcomers enjoyed the ability to "walk in safety in the midst of enemies who in the hottest fight lay down their weapons when it is displayed." It allowed them to overcome both the linguistic differences and the problematic circumstance of having no shared kin, transforming outsiders into kinsmen and "naturalizing [them] as such." As French observers noted, the calumet was the passport—the "pass and safe conduct" mechanism—for diverse people who did not share kin to get along in the world. For rival groups looking for a "glue" capable of keeping their divisions at bay, the calumet had important work to do.[46]

Calumet ceremonialism did not iron out all the rivalries. It could not, and despite naive assumptions on the part of French observers who considered the calumet a magic talisman of peace, it really was not meant to do so. To the contrary, an important function of the calumet was to create unity by expressing

"Captaine de la Illinois Nation," with his calumet and tattoos, which the French called *mataché*. From Louis Nicolas, *Codex Canadensis*, ca. 1700. Courtesy Gilcrease Museum, Tulsa, Oklahoma.

common enemies, that is, both to adopt *and* to exclude. And to do this, participants frequently employed a special kind of symbolism. Welcoming Jesuit Jacques Marquette to their village in 1673, for instance, the Peorias (Illinois) danced the calumet, impressing Marquette so thoroughly that he spent many pages describing his experience. Central to the whole ceremony, Marquette understood, was the violent symbolism of captive-taking. Not only did the dancers recount "all the captives they had made" while singing their songs, but the chief of the village ended the ceremony by actually gifting Marquette "a little Slave." As he described it, this gift was the ultimate bond: a bond of kinship. "Here is my son," Marquette recalled the chief saying, "whom I give you to Show thee my Heart."[47] It was typical. As several eyewitness accounts suggest, one of the most important symbolic acts featured in calumet ceremonialism was the gifting of captives—one group giving to another group an enslaved outsider they had subdued through mourning war.[48]

Marquette and many other Frenchmen frequently described the calumet ceremony as inscrutable and even "altogether mysterious," but in truth it is not difficult to understand the meaning of the captive within calumet ceremonialism. The gifting of a captive achieved two things. The first one is already familiar: through the symbolism of gifted captives, strangers could express their mutual and reciprocal obligations and make ties based on shared affection for a lost kinsman. In this way, one logic of gifting captives in a calumet ceremony was always to create solidarity among two groups or lineages—"*I bring you my flesh*" (emphasis added).[49] But the second function of the symbolism of the gifted captive was even more important, because the meaning of a gifted captive was not only defining the relationship between the giver and the receiver; it was also defining the relationship between both parties and the *captive's* kin. Very clearly, passing a captive implied that both participants in the transaction agreed to consider the original kin of that captive as their enemy. When the Peoria chief welcomed Marquette, using the slave to "show my Heart," the chief not only expressed his affection and open-heartedness toward Marquette, but his utter disdain for the group from which the captive had originally come. And, importantly, by receiving the captive, Marquette also showed his own heart, implicating himself and joining the Peoria in their act of hatred. The strategic power of this symbolism should not be underestimated. Captive-giving and captive-*receiving* created a clear symbolic social wedge meant to cleave apart allies (fictive kin) from enemies (nonkin). Even when an actual gifted captive was not part of the ritual, the pageantry nevertheless featured a recounting of the dancers' battlefield achievements and especially their histories of captive-taking and mourning wars. Naming their enemies and these conflicts was at the heart of defining new friendships. The complex symbolism of calumet ceremonialism was an expression of peace and fictive kinship but also a drama about warfare and enemies—naming the people who were not kin.[50]

Given the social dynamics of newcomers and strangers living together, this logic of captive-exchange became especially significant to intergroup relations in the ecotone. In a landscape of strangers, captives were first of all needed to create and cement fictive kin and define alliance. But in a landscape of strangers, potential captives—those who did not share kin with an ally—were also all over the place.[51] Algonquians like the Meskwakis and the Illinois, for instance, shared no kin with Siouan-speakers like the Ho-Chunks, so they were "available" targets for captivity. Meanwhile, and more importantly, just to the west of the prairie ecotone was the beginning of another social milieu comprised of groups even further remote in the eastern plains—groups like the Ioways, the Quapaws, the Osages, the Padoukés, and the Pawnees. Living "well night the end of the world" put the ecotone residents in reach of these *utter* strangers, who could be gifted and traded in order to cement alliances and build kinship networks. And so need and availability merged to create the special function of captive-raiding in the ecotone. As Antoine Denis Raudot put it, this was the heart of diplomacy: "Usually [captives] are used to replace the dead, but often some are also given to other nations to *oblige these nations to become their allies*" (emphasis added).[52] Just as captives could symbolize reciprocity among individuals and families within the same villages, the gift of a captive's life could create a powerful reciprocal obligation among the members of different villages.

Diplomacy based on captive exchange followed two main strategic calculations. In a bilateral situation, captors could return a captive to his kinsmen in order to end war, restore peace, or solidify alliance. Accepting returned slaves from one's own village meant ending conflict, if not also accepting a measure of humiliation. Sending them home reflected the captors' "magnanimity." In the 1660s the Sioux developed a reputation for this kind of strategy when they repatriated many captives belonging to enemies who had trespassed in their territory. Such acts were surely a diplomatic performance, a display of "glory," not only for the kinsmen of the repatriated captives but to others who witnessed the "justice"—and the power—that the Sioux exercised toward their neighbors. As Raudot put it about the Sioux in particular, this repatriation was their general strategy: "They generally send back any prisoners they make, in hope of obtaining peace." Sometime after the attack on the Ho-Chunks described earlier, the Illinois evidently sent back a good number of the captives they had taken, likely for this diplomatic purpose as well.[53]

A second and more complex calculation underlying captive exchange was to pass a slave that was unrelated to either group. This was what the Potawatomis attempted to do when they gave to the Miamis an Iroquois slave to encourage a new alliance, and what the Miamis signaled when they gifted a group of Iroquois prisoners to the Meskwakis. The idea here was not just to solidify friendship, but to define hatred for a shared enemy. This function was especially important when a group passed a slave to a *new* neighbor, a group of strangers, as when the Iroquois gifted two slaves—from the Ottawas and the Shawnees, respectively—to the Sulpician priest François Dollier de Casson in 1669. Indeed, the French had a good deal of experience with this. As they would realize over time, gifted slaves were frequently all about implicating them in the animosities that certain groups felt for each other. The idea was not just to "oblige people to accept the captives," as Raudot put it, but also to oblige them to accept the animosity for the prisoner's kin that the captive's status symbolized.[54]

Given both of these strategic logics on the part of the giver, it is necessary to note the strategic calculus that gifted captives created for the receivers in these transactions. Since the gift of a slave clearly implicated the receiver in the giver's conflict, the receiver had important decisions to make too. The receiver could refuse the captive, or even return a captive to his kinsmen. Importantly, this is what the Meskwakis did when the Miamis offered them gifts of Iroquois prisoners, turning around and offering to repatriate these captives soon afterward.[55] Rejecting the Miamis' idea of an anti-Iroquois alliance resting on the symbolism of these slaves, the Meskwakis asserted an important alternative vision, showing their "magnanimity" instead to the Iroquois. If groups like the Ioways aimed to stay neutral in conflicts between groups, they had to make such calculations when their neighbors offered "a little flesh," that is, a slave.

In the contact period many French hardly understood the complexities of this diplomacy. They initially understood slaves as innocent "gifts." Yet much evidence suggests that much more than "peace" was intended when the Ottawas gifted a slave to Marquette in the 1660s and one to Dulhut in 1678, in effect pressing the Frenchmen to join them in defining their enemies as

French enemies also. Moreover, not only did the calumet express violence and animosity as well as peace, it also came with an expectation of action and it was inviolable. In short, the calumet—and the culture of captivity that underlay it—was viewed as the source of order in the ecotone, the source of "everlasting" agreements and relationships.[56] If the downside of living in a borderland of strangers was a vulnerability to captive-raiding, the upside—especially for groups looking to assert power—was the ready availability of strangers, people who could be easily transformed into the most powerful symbols for building and demonstrating alliance and division.

◆ ◆ ◆

Patterns of violence and peacemaking based in captive exchange defined the volatile order that grew in the ecotone region in the mid-1600s, just as the contact era was opening up. Of course, with the beginning of the Beaver Wars—the transformative conflict among Great Lakes Indians and European trade allies that brought Iroquois attacks like the one in 1680—outside forces only complicated and interacted with these ongoing patterns, producing more rivalries, more demand for captive-based diplomacy, and more newly migrated "strangers" to supply captives and negotiate meaning. But previous and ongoing trajectories in the ecotone would always be as important to defining this region's history as invasion. Understanding how this played out is necessarily difficult, owing to imperfect sources. Nevertheless, we can identify at least three broad patterns that emerged out of these local trajectories.

First, we know that through these diplomatic engagements ecotone people made some broad and relatively coherent "coalitions" and "affiliations."[57] In broad strokes, there were five main clusters: Miami-Illinois, Fox-Sauk-Kickapoos, Sioux, Cree-Monsoni-Assiniboines, and Ojibwa-Ottawa-Potawatomis.[58] To be sure, despite the idea that the calumet was "inviolable," in practice it is clear that these coalitions were often loose and temporary. The "hodgepodge" of social relations made shifting alliances a part of life in the ecotone. Consider, for instance, a village like the large settlement near Green Bay that in the 1660s was occupied by a motley group of Meskwakis, Sauks, Mascoutens, Potawatomis, Kickapoos, Miamis, Ottawas, and Hurons. Nearby lived groups like the Menominees and the Ho-Chunks. Importantly, these latter two groups put aside their hostilities to several of the other inhabitants and even accepted the Illinois—perpetrators of that massive attack against the Ho-Chunks in the 1640s—when they arrived on a sojourn. Intergroup relations were a matter of reasonably stable coalitions but also frequent realignments, including temporary armistices in this "babylon of tribes and dialects."[59]

Second, and central to the ways in which these coalitions and armistices took shape, one group in the ecotone emerged with a special strategy: the Illinois. In the 1660s, a generation after their fateful encounter with the Ho-Chunks, evidence suggests most of the Illinois speakers lived west of the Mississippi River, where they frequently fought with the Sioux, the Meskwakis, and occasionally with the Iroquois. Attracted to fur trade opportunities when the French first arrived, however, the Illinois initiated a pattern of aggressive diplomacy designed to erase neighbors' animosity toward them and to strike a regional peace. The way they did it was crucial. Sometime in the mid-seventeenth century the Illinois began raiding to the south and west of their location near the Des Moines and Illinois Valleys. Their targets were distant Siouan and Caddoan peoples like the Pawnees. The French noted "warlike" behavior among the Illinois when they first arrived in Illinois Country in 1670s, observing how they brought captives from these raids for "trade" to northern groups like the Ottawas and the Sioux, at trade and mission centers like La Pointe on Lake Superior. As Marquette put it, "They are warlike, and make themselves dreaded by the Distant tribes to the south and west, whither they go to procure Slaves; these they barter, selling them at a high price to other Nations, in exchange for other Wares."[60]

What the French saw as "trade" and "barter" did surely have an economic logic. Yet the meaning of these captives and "slaves" was more complex. Kinless captives were valuable in the ecotone, owing to their symbolic value in both mourning war and diplomacy. Thanks to their proximity to distant Siouan and Caddoan people like the Pawnees, the Illinois soon had a special role: supplying these needed captives to neighbors in the Great Lakes. As Jesuit Claude Allouez noted, "The Ilinois are warriors and take a great many Slaves, whom they trade with the Outaouaks for Muskets, Powder, Kettles, Hatchets and Knives." Making war "against 7 or 8 different [Western] nations," the Illinois made peace with the eastern Sioux, recently their bitter enemies, and the Ho-Chunks. Moving to a large and concentrated village at the top of the Illinois Valley, the Illinois built their own center of power in the ecotone region, welcoming—if not subduing—a diverse group that soon included Shawnees, Mascoutens, and others.[61] Here the French soon promised to create a whole new trading outpost.

The Illinois were not the only group attempting to use their captives strategically to shape alliances in this period. Groups like the Miamis were doing it as well, for instance, when they gifted a group of Iroquois captives to the Meskwakis around 1672. And this brings us to one final pattern of intergroup relations that emerged in the ecotone in this period, which has to do with this latter group—the Meskwakis. Receiving these Iroquois captives from the Miamis, the Meskwakis did something peculiar: they reached out to the Seneca relatives of the prisoners and offered to repatriate the captives.[62] Even disregarding a recent raid that the Iroquois had made on a Meskwaki village, the Meskwakis sent a message through the Jesuit Claude Allouez that they wanted to forget about all that and forge a new alliance between themselves and the powerhouse to the east. "When thou seest the Iroquois, tell them that they have taken me for some one else. I do not make war on them, I have not eaten their people; but my neighbors took them prisoners and made me a present of them; I adopted them, and they are living here as my children."[63] Offering to repatriate the Iroquois captives gifted by the Miamis was a starkly surprising strategy. French observers wondered why the Meskwakis did not "remember all the treacheries of the Iroquois" who had caused "the destruction of a great many of their finest young men" in numerous raids on Meskwaki villages in recent years, not to mention on their neighbors.[64] But the Meskwakis were evidently calculating boldly. Previously the Meskwakis had sent captives to the Iroquois, such as when they delivered a group of Illinois to the Senecas in 1667. In so doing they were rejecting solidarity with their ecotone neighbors and holding open the possibility of alliance with the Iroquois. Importantly, the act seemed specifically designed to counterbalance the Illinois' own power, since now the Iroquois promised to focus their raids against the Shawnees and the Illinois themselves. One hint of the Meskwakis' underlying logic lies in what they told Father Louis Hennepin when he traveled through a Meskwaki village in 1679. Warning them not to visit the Illinois, the Meskwakis revealed their fear that the "Illinois should grow too powerful."[65]

◆ ◆ ◆

The fact that in 1679 the Meskwakis feared Illinois power is telling and important to acknowledge. It does not fit with historians' usual accounts of what was happening in the ecotone region at this time however. Ignoring the Meskwakis' assessment of intergroup dynamics in the region, historians have normally painted a picture of a region *invaded*, its people pushed by exogenous forces to the verge of desperation, not power. Historians have especially emphasized the Illinois' victimhood, exaggerating and misunderstanding the events of the following year—1680—and the Iroquois raid in which so many Illinois were killed and taken captive.

A better way to understand this period is to listen to the Meskwakis themselves and to notice what was actually shaping prairie Indians' history. Just as the Iroquois were arriving for the fateful 1680 attack, a massive army of the Illinois was setting off under the leadership of the chief, Paessa, to raid some unknown village for captives. Paessa's raid not only has everything to do with why the Iroquois were able to strike such a major blow against the Illinois in this period, but it was also at the heart of an important regional history. In this period, diverse ecotone residents used captives to shape tentative coalitions in the midst of a volatile borderland, creating a regional order based on Indigenous logics of mourning war and reciprocity. These ongoing local trajectories of migration and relationship-building shaped history as much as any invasion in this period.

It was a volatile diplomacy, but nothing "mysterious," as the French often complained in their observations of the calumet. In the traditional account of contact-era history in the Great Lakes, supposedly disordered intergroup relations among Indigenous people called for French intervention, for it was only through French mediation that tensions among constantly antagonistic groups could be resolved. But this version of history does more than reify a heroic self-image through which French colonists viewed themselves as benevolent mediators. It also ignores the logics by which Indigenous people created their own order in their relationships, despite—and in some ways based on—the region's volatility. In the ecotone, *captives* were the *Indigenous glue* for coalition-building and alliance-making, the symbols by which diverse peoples—and particularly the enterprising Illinois and independent Meskwakis—invented a regional order that fit their own histories and their own goals in a period of transformation.

If captive-raiding and exchange were the key to the Native-centered history of the ecotone in this period, however, these were clearly complex strategies. The phenomenon of captive-trading was not just about settling conflicts. It was also about causing them and exploiting them. Pursuing their ambitious strategies, the Illinois began to gather in ever-larger villages for security, as they increasingly did in the Illinois' Grand Village in the 1660s and 1670s, located at the top of the Illinois Valley. When Jesuits first came to the Illinois

Country in the 1660s they noted this population center ballooning. It would eventually become a massive settlement, which at its height contained probably 12,000 Illinois speakers and many of their Algonquian allies.

It was here that the history of the Illinois and the ecotone region in general would enter a new phase. We have so far explored the history of intergroup relations in the ecotone region as mostly a *human* story. But in large villages like the newly forming Grand Village of the Kaskaskias, the strategy of captive-raiding and the forces of colonialism were about to entangle with the special nonhuman conditions of the ecotone in important ways. Conspicuously, the location of the Illinois' largest settlement at the top of the Illinois Valley was in the heart of a major bison habitat, and probably the best in the region. In the 1680s and 1690s, bison and other nonhuman actors would play an ever-more-central role in shaping the history of the ecotone region and its human conflicts.

5

THE GREAT BISON ACCELERATION

TRAVELING AND LIVING IN THE ILLINOIS COUNTRY IN THE 1680S AND 1690s, French eyewitnesses wrote some of the richest and most voluminous descriptions of any North American Indigenous group of the colonial period. And although priests like Jacques Gravier and officials like Pierre Charles de Liette wrote broadly about the spirituality, economy, and technology of the Illinois, their accounts almost always returned to what they saw as a most peculiar feature of Indigenous social life in this moment: the interactions between men and women. For instance, observing what they understood to be deep tensions in the Illinois culture, the priests and explorers noted a few cases of outright physical violence by Illinois men against women. But more consistently they noted the strict and unequal way in which men and women divided their spheres, especially when it came to work. To the French one of the most conspicuous things about Illinois culture was that many women worked all the time. As Jesuit Sébastien Râle, the famous missionary to the Micmacs who spent two formative years in Illinois in the 1690s before taking up his post in colonial Maine, put it, "women work from morning until evening like slaves."[1]

Meanwhile, French eyewitnesses observed that most men's lives were quite different. While women worked obsessively inside the village, men did short bursts of intense work while hunting, and then no more. To some eyewitnesses like Gravier and Julien Binneteau, Illinois men were simply lazy.[2] But others perceived that men were just busying themselves with completely different activities. Conspicuously, if women were occupied with all things related to the domestic and village economy, men were obsessed with something else: preparing for and conducting war. Râle remembered Illinois war parties routinely embarking on journeys of hundreds of miles "through the midst of forests in order to capture a slave." For Illinois men there were few activities more meaningful. "They count as nothing the hardships" required by their long-range captive raids, Râle wrote. Along with hunting, slave-raiding success was where Illinois men were able to "make their merit consist, and it is this which they call being truly a man."[3]

To the French this seemingly stark division of labor—men out fighting while women toiled away in the village—shone an especially harsh light on Indigenous masculinity and its seemingly shallow priorities of "glory"-seeking.[4] As the French saw it, women's work fueled an extraordinary type of masculine freedom and by extension the kind of excessive and wanton militarism that French eyewitnesses thought depraved the men with vanity, cruelty, and violence. Far from honorable manhood, they thought the Illinois social order enabled degeneracy and dishonor.[5]

These criticisms of the gender order among the Illinois were part of a larger European stereotype about Indigenous people, and thus somewhat generic and useless.[6] Yet they were also specific to the Illinois, including idiosyncratic details about the way in which husbands and wives interacted, especially in the polygamous families that defined Illinois households. Given their specificity, these descriptions have intrigued historians, who have long interpreted them earnestly in the context of culture change and especially religious conversion. Tensions like this explain why Illinois women enthusiastically embraced teachings by Jesuit priests, for instance, and why they sometimes married French fur traders—in order to "resist" these circumstances.[7]

While surely historians are right to focus on these important *results* of the gender order in Illinois, however, they have usually ignored evidence about its possible *causes*. Eyewitnesses often misunderstood Illinois gender relations, as I will show, but their accounts do reveal a complex hidden logic of these arrangements, especially when it comes to how the burdens of women's work fit together with the extremes of male militarism. The French were right to suggest that the two were closely related in this period. But if the French believed that female toil underwrote and fueled male soldier culture, in fact the opposite may also have been true. To understand why we need to explore evidence about *who* many of these hardworking women in Illinois villages were, and *what* they were doing with all of their labor. To begin this exploration we need to return to what was the most important material reality in the Illinois lifeway: the bison. By the 1690s bison hunting had been shaping life and community for prairie people for generations. But now the bison economy had shifted into high gear as the Illinois doubled down on bison hunting to supply needed calories to their huge village at the top of the Illinois Valley. Settling in an area where bison proliferated in huge numbers, the Illinois used that extraordinary energy source to power a massive population center, one of the largest in North America at the time. Meanwhile, it was here in the early 1680s that Robert La Salle opened a market for trade in bison hides, which his successors La Forest and Liette continued at Fort St. Louis. In response to both of these demands, production in the bison economy skyrocketed. Describing a great acceleration in bison hunting in the eastern prairies, Râle wrote that the Illinois never took fewer than two thousand animals per year in this period. In one single five-week outing around 1687, Liette observed the Illinois killing twelve hundred animals.[8]

Consider what this acceleration in bison exploitation meant in terms of work, however. The ordinary bottleneck of the bison economy was always the processing stage. It was a huge and messy job of work, now made more intense and burdensome by the massive scale of the Illinois' expanding economy. Two thousand dead bison meant potentially well over a million pounds of usable meat for the thousands of people at the Grand Village, but only if it could be cured and stored quickly. Hide tanning, meanwhile, was even more demanding and laborious. Importantly, these parts of the bison economy were always women's work.

To preserve community—the egalitarianism and collective spirit that might be thought of as the "software" of bison hunting—in this booming economy was a challenge. The scale of bison hunting in these years conspicuously put a much heavier burden on women than on men. How to preserve gender "complementarity"—the balance between different but equally valued work roles for men and women—in this scenario? The Illinois and other prairie people had a solution ready at hand. At the Grand Village, mourning war took on a new meaning—and a new purpose—as captives became valued just as much for their work as for their role in kinship replacement and diplomacy. Increased bison production accelerated the demand for captives, as men went on raids not just for glory (as the French thought) but also to obtain the laborers that were the key to the competitive Illinois economy and the preservation of the community. Here, then, was the hidden material logic of the gender order in Illinois villages in the 1690s, when French eyewitnesses saw oppressed women "resisting" their status and seeking refuge in Christianity. Many of the women who Sébastien Râle witnessed working "from morning until evening like slaves" likely *were* slaves, as were other women in the Illinois village whom Frenchmen viewed as oppressed and degraded.

Although the French witnesses may not have realized it, bison hunting and its special demands helped shape the gender dynamics among the Illinois they considered so peculiar. At home in the village, polygamous families were the central manifestation. The logic was simple: degraded processing work was "insourced" to captives who became second and third wives, freeing native-born Illinois women to be exalted as mistresses of the household. Militarism and bison hunting were cause and effect of each other, mutually constituted aspects of a peculiar kind of Indigenous power rooted in the special dynamics of the ecotone.

Yet just like that special geographic region, this entire structure was unstable. The strategy rested on a volatile ecology and risky social patterns. In a series of dramatic turns, the ecology of the region shifted (as it was prone to do), while intergroup tensions removed constraints on overhunting. When environmental conditions brought the Grand Village to a fast end, however, it was not a simple case of environmental determinism. Instead, nonhuman environmental changes proved consequential in this moment precisely because all the key features of Illinois life—large villages, polygamous families, militarism, colonial trade, and regional power—were entangled together in cycles of cause and effect, with the bison a fundamental and active participant at every turn. In this *collectif,* human and nonhuman agents worked together in co-constituting fashion. This coalition of complex agencies "held together" only because the bison in the ecotone—the herds of five thousand animals that Râle observed "as far as the eye could see" in the early 1690s—briefly squeezed into the upper Illinois Valley in unprecedented numbers.[9] When climate change and overhunting cut into the herds, it was the relationality of human and nonhuman agents that made history.

◆ ◆ ◆

In the wake of the Iroquois attacks of 1680, the Illinois speakers united in one of the largest population centers east of the Mississippi at this time. At the top of the Illinois Valley, near modern-day Starved Rock, a settlement known as the Grand Village of the Kaskaskias swelled with people, including the bulk of well over ten thousand Illinois speakers and certain other related groups. It was an unprecedented arrangement.[10] Some Algonquian speakers, like the Shawnees and the Mascoutens, also sojourned to near the main village for brief periods. But the Illinois speakers themselves lived there together for most of a generation in a massive permanent summertime settlement, before moving a short distance to the region near modern-day Peoria in a second village called Pimitéoui. At both of these settlements French interlopers lived alongside the Illinois, leaving several rich written descriptions. Together with archaeology excavations from the areas, including major excavations from what archaeologists term the Zimmerman site, the written record makes it possible to deduce much about what was going on as the Illinois united in these huge villages.

As eyewitnesses and later historians have surmised, congregating in these new villages likely emanated from a defensive logic, as a large village population translated into military power to resist attacks by Iroquois and other hostile neighbors. Additionally, by the 1680s part of the logic for the congregation in the Grand Village likely also had to do with the presence of the French. The Jesuits first became a magnet when they set up their mission in Illinois in 1673, even though Marquette and Allouez initially spent just a few months each in the village. In 1689 Jacques Gravier returned and put the mission on a much more stable footing. Meanwhile, more important than the Jesuits were the traders. La Salle chose the area of Peoria for the location of his Fort Crèvecoeur, and later established Fort St. Louis right next to the Grand Village; both of these were to become trading centers. If the French initially chose these sites in part for their proximity to the Illinois' settlements, they had the reciprocal effect of increasing the Indigenous population in the nearby villages. All these reasons explain why the villages swelled with people.

If these considerations—the need for defense and the presence of the French—were part of the motivation behind the size of the massive community at the Grand Village, there was yet another underlying and fundamental factor shaping not just its size but also its precise location: environment. As described earlier (see chapter 2), in the 1500s and 1600s the whole region of the ecotone had recently experienced a species shift, with bison "squeezing" into the eastern prairies in herds for the first time. But something special was happening in the area around the upper Illinois Valley in this moment.

Though the bison population of the eastern prairie normally fluctuated, sometimes growing large and sometimes "growing somewhat scarce" from year to year, in the late 1680s and 1690s, and especially around the Grand Village, the herds were enormous.[11]

Consider, for instance, Sébastien Râle's impressions of the special situation. Recalling his arrival at the Grand Village as a young priest, Râle swooned over the local ecotone environment as a whole, the most prosperous in North America, as he figured it. He thrilled at the amazing wetlands prolific with flocks of birds and the Illinois River abundant with fish, but it was the prairie ecosystem surrounding Grand Village—especially the bison—that really awed him. The animals were "countless," so prolific that "when [hunters] have killed a [bison] that seems to them too lean, they are satisfied to take its tongue and go in search of one that is more fat."[12] Râle recalled a moment when the herds were so plentiful that they seemed inexhaustible, an impression that Hennepin shared in a slightly earlier period when he noted that the Illinois would be "unable" to destroy the herds in this location no matter what they did.[13] As we will see, Hennepin's perception that the herds were indestructible was false. But the French were clearly correct that a bison boom was centered on the Illinois Valley; the largest herds seem to have been located right at the location of the Grand Village and the prairies immediately surrounding it.[14] As seminary priest Jean St. Cosme observed when he entered the Illinois Valley from the Kankakee, bison were everywhere. "On that day we began to see oxen," he wrote, before embarking down the entire river. "We afterward saw some nearly every day during our journey as far as the Acansas."[15] It was a special time.

It is important to understand how the presence of these huge herds related to the defensive strategy that the Illinois pursued by congregating in the Grand Village. Large villages like this were common in the *pays d'en haut* in the period of the Beaver Wars as a response to the attacks that Iroquois soldiers launched frequently through the 1670s and 1680s. But at the Grand Village the bison gave the Illinois a special advantage. For other groups in the Great Lakes, the defensive benefits of large agricultural villages lasted only through the summertime. In the wintertime many groups became more vulnerable as they split into much smaller hunting parties in order to stalk solitary animals like deer or even small herds of bison. At the Grand Village the requirements and affordances of hunting the massive herds meant that nearly the whole population of the Illinois could—and indeed had to—stay together all year. Allouez could see how important this was. In the heart of the prairie ecotone, "[bison], by their abundance, furnish adequate provision for whole Villages, which therefore *are not obliged to scatter by families* during their hunting season, as is the case with the [Indians] elsewhere" (emphasis added).[16]

This was an unusual situation, and the French commented on it frequently.[17] Nicolas Perrot was even more explicit about this strategic "side effect" of hunting the huge herds. Because of the large hunting parties involved in the bison economy, "they are always ready and able to defend their families against their enemies; for the families are always on the march, placed on the flanks which are protected on the right and left by the warriors, and sheltered from the attacks that might be made on them." While on a bison hunt "it is impossible, therefore, for the enemy to appear without the entire troop knowing it." It was as if the defensive logic of the large population in the upper Illinois Valley synchronized perfectly with the demands and affordances of bison hunting. As Perrot summed it up: "The people of an entire village go together to this [bison] hunting . . . for two reasons: the first, in order to defend themselves against the attacks which their enemies might make against them; and the other, that thus they may be able to drive in a greater number of animals."[18]

If the defensive advantages of bison hunting helped lead to unprecedented population concentration in the Grand Village in the 1680s and 1690s, bison were also clearly central to another significant logic of the huge village: trade. This too is a more complex story than it appears on its face. Of course, French traders began traveling to the distant pays d'en haut in the 1660s and 1670s, and the earliest fur traders likely arrived in the Illinois Country around this same time. Meanwhile, Illinois speakers themselves traveled north to established trading centers like Saint Esprit on Lake Superior in order to initiate trade as early as the 1660s. Here French traders ordinarily sought beaver, the main commodity of the Great Lakes fur trade in the colonial era. And yet for the Illinois this was a problem. Surely there were beavers present throughout the streams and wetlands of their rich region, especially around the Illinois River. But as many sources make clear, the French market prized beaver pelts from more northern climates, where cold weather encouraged thicker, denser, and more desirable fur in the animals. The Illinois' beaver pelt was, by comparison, inferior. For example, Antoine Denis Raudot observed that "all the pelts which come from this region [of the Illinois], which is the south, are not as esteemed as those from the north, not being well enough furnished with hair."[19]

So it was logical that the Illinois might turn to another commodity for trade: bison. It was just a fledgling enterprise, but a promising one. French observers fantasized about how bison could be exploited and marketed. There was the prospect of domesticating the animals and raising them for meat and milk or for use as draft animals. Their wool could be exported. But to Robert de La Salle, the main chance was in another product of the animals: their hides. In the original patent La Salle obtained for his colonial enterprise in 1678, the explorer secured a monopoly on the bison trade. Writing of his impression that bison in Illinois were "an almost inexhaustible treasure here," La Salle wrote that "the multitude of oxen is beyond all belief. I have seen twelve hundred of them killed in eight days by a single band of Indians." Importantly, he laid out a vision for a future large-scale enterprise among the Illinois, a tanning operation: "If we had the tanners, we could not only supply [bison hides] to Canada but could send them to the [Caribbean] Islands and even to France." As he told courtier Abbé Claude Bernou, the result would be a "a large profit if the hides were worth 25 livres ready dressed." The hide trade in Illinois opened in 1683, when La Salle ordered his men at Fort St. Louis to "get together as many buffalo skins as you can . . . and encourage the Indians to kill [bison]."[20] The Illinois soon began supplying bison skins to the French in large numbers.

Bison thus underlay the consolidation of the massive population at the Grand Village, constituting the often-ignored material basis of the defensive strategy at the village as well as the Illinois' trade relationships. At the same time, building their lifeway around these benefits presented the Illinois with new imperatives in terms of social life and culture, imperatives that went well beyond the hardware and software of pedestrian bison hunting we have already explored as a component of prairie Indians' ethnogenesis. For if bison hunting itself was already well-established, the large scale of bison production was new and unprecedented, and it created new challenges. The biggest issue was labor. It's time now to see how the Illinois met the new demand for labor as they accelerated the bison economy, to dig into the actual material reality of what it meant to kill and process hundreds of these animals at a time.

◆ ◆ ◆

As the population of the Grand Village grew, and as bison hunts got bigger, the work of the hunting increased overall. But that labor increased especially within certain dimensions. Considering the whole task of bison production from start to finish, the most important limits on production in the bison economy were never found in the actual killing of animals. Regardless of how many animals the hunters wanted to take, once the work of running down and trapping the animals at the kill site was accomplished, it cost hunters relatively little in additional marginal labor to yield each additional animal. But in processing unlike in hunting, each additional animal added a proportionately large amount of labor to the total. In other words, processing the bison kill was definitely the bottleneck in the whole system of production.[21] Just like the bison hunters themselves did, we should slow down here to consider this extraordinary material reality as revealed in various eyewitness accounts from the ecotone region of the time.

Descriptions of bison processing in the field make it clear that it was messy, laborious, and intense. There were several stages. Bison were enormous animals, and, depending on where they were killed, they sometimes first had to be moved some distance to a processing site. Witnessing his first-ever bison kill in the field, Marquette reported in 1673 that it cost "three persons . . . much difficulty" to move what was clearly a small bison. For Hennepin a single bison required *thirteen* men, who also had "much ado to drag him from the Place where he fell."[22] And this was, of course, just the start of many days—if not weeks—of work.

The next phase in processing—butchering—was a spectacle of appropriate technology and straight-up toil. The butchering never failed to impress observers. Hennepin felt "no small matter of Admiration to see these [Indians] slay the Bull, and get it in pieces; they had neither Knives nor Hatchets . . . and yet they did it dexterously with the Point of their Arrows, which was made of a Sharp Stone." The butchers worked for hours on end, cutting the huge animals into usable pieces, in both the heat of the summer and the cold of the winter.[23]

Next came the actual preservation, the most resource-intensive aspect of the process. Eyewitnesses marveled that prairie people had developed an impressive technique that could cure meat in such a way that it would last as long as four months without spoiling. First, they built the principal tool of the processing camp: a *gris*, or large wooden "gridiron" over which they laid the heavy animal carcass to dry. Sometimes they could accomplish the task of curing with salt and the heat of the sun, but it seems they usually used smoke to cure and preserve the meat. This of course meant gathering wood, not only for building the drying racks but also for fuel to make the low

Indians smoking their meat. Detail from Louis Armand de Lom d'Arce Lahontan, *Nouveaux voyages . . . dans l'Amérique septentrionale* (1709). Image courtesy Rare Book & Manuscript Library, University of Illinois at Urbana-Champaign.

smoldering fire over which they smoked the meat for long periods. To meet both demands, wood had to be cut and dragged from the forest edge to the field camp, which was no small task. Even in the best of circumstances, when the kill was made in the ideal location relative to timber resources, there were always many, many loads of wood to carry.[24]

The task of preservation itself could consume a week, with workers bustling through busy camps. Workers tended constant fires and struggled to move hundreds of pounds of meat through the various stages of preservation, though this was not even close to the end of the process. Since prairie people also utilized many other parts of the animals, including hides, wool, and bones for technology, the work of bison processing continued long after the meat—featuring especially the *plat côté*, or ribs—was finished. As one eyewitness to a summer hunt estimated, bison processors worked in the bison hunting camps for a full three months after the hunt was over to finish all of the accompanying work, from spinning wool to tanning hides. While some hunts yielded modest numbers of animals, it seems clear that this three-month estimate could sometimes lengthen. Imagine how much processing labor must have been created in cases like the hunt that Perrot witnessed, when a single day in the field yielded fifteen hundred animals![25]

After all this work there was one final and major task: transporting and delivering the finished products of prepared bison meat, hides, wool, and tools back to the permanent village. This may have been the most labor-intensive part of the entire weeks-long process. Like the butchering, European eyewitnesses marveled at the skill involved, as well as the tremendous strength, as the hunters moved the loads. They especially noted the amazing weight that each member of the village could carry over long distances. Contrary to myths, the bison hunters rarely used the whole animal—they simply could not manage to move it all, so inevitably much was wasted. But they did their best, not only by cleaning the bones of all the meat that they could manage while in the field—and eating first only those parts that could not be preserved—but also by carrying back as much as they could.[26] This resulted in one of the most impressive feats in all of the physically demanding work of bison hunting. According to Liette, individuals in the Illinois hunting party each carried the cured ribs of eight animals all day over the course of an eighty-mile journey.[27]

There is much more we could observe about these practices, especially the ingenuity involved and the difficulty of the job. But there is one fact that stands out above all else as we observe their historical significance: a disproportionate amount of this work was performed by women. The huge jobs of processing the animals—from dragging the recently killed carcasses to hauling the "flat sides" of meat to the village—were female-gendered work among the prairie people. As Raudot put it: "It is the women who have the care of gathering this meat and smoking and preserving it."[28] Hennepin echoed Raudot, noting that after the animals were killed, the men "come to notify their women, who at once proceed to bring in the meat." Illinois women not only smoked but then also transported the meat, with this latter part of the labor standing out especially to European observers: "Some of them at times take on their backs three hundred pounds weight, and also throw their children on top of their load which does not seem to burthen them more than a soldier's sword at his side." To Hennepin the women were "so strong, that few Men in Europe can match them; they'll carry Packs that two or three can hardly lift up."[29] Hennepin's comparison here again employs a stark European stereotype of Indian women, not to mention a dim and prejudiced view of Indigenous masculinity. But however Europeans may have viewed it, this

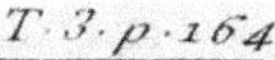

Indians of the North [upper Louisiana] *who go hunting in the winter with their families.* The Illinois' bison economy demanded many workers to process hides and preserve meat. From Le Page du Pratz, *Histoire de la Louisiane* (Paris, 1758). Image courtesy Rare Book & Manuscript Library, University of Illinois at Urbana-Champaign.

division of labor was clearly strict: men killed the animals, women were responsible for processing and transporting the meat.

If meat processing was women's work, other aspects of the bison economy were their domain too. Prairie women developed a whole technology for brain-tanning the tough bison leather and making it, as one Jesuit remarked, "as supple as our chamois skins dressed with oil." Prairie women also made clothing, as well as tools, from bison leather and bones. Women were responsible for "working up the hair of the oxen and . . . making it into leggings, girdles, and bags." Perrot noted that "it is for [the women] to dress the skins of the animals which the husband kills in hunting, and to make robes of those which have fur." Eventually these crafts, as well as the porcupine quillwork that women added to their hides, would make prairie artisans famous in Paris in the eighteenth century. As another observer wrote, the overall division of labor was clear: "Hunting and war form the whole occupation of the men; the rest of the work belongs to the women and the girls."[30]

It was a stark reality. To be sure, it was not atypical for Algonquians to divide labor by gender, but among prairie people it seems the unusual demands of bison production increased the women's share of labor, or what was meant by "the rest of the work." This was especially true as the Grand Village increased in size, which maximized bison production in the 1680s. Many sources suggest that these patterns disrupted earlier arrangements. Invading the prairies in the early 1600s, the prairie people had become "societies with bison" but not fully "bison societies." They did not abandon agriculture, and their annual food-production cycle continued to include other diverse sources. In the spring Illinois women farmed and grew crops, particularly corn at the Grand Village. Illinois women also gathered a good deal of food, such as tubers and berries, from the ecotone's marshes and streams. These parts of the economy traditionally were part of the "complementary" division of labor that characterized many prairie groups. Controlling the agricultural production exclusively while adding their share of labor to a modest-size hunting economy, women were traditionally not overly burdened and their labor was a source of control and power. Creators of life and equal partners in the provisioning of food, women added their labor in a complementary and balanced division of labor.[31]

But evidence suggests that large-scale bison hunting skewed the balance, especially starting in the 1680s. Women faced a massive increase in the number of people to support and were "continually occupied" and "humbled by work."[32] As Râle put it, the bison economy put a large labor burden on the women in particular: "As for the women, they work from morning until evening like slaves. It is they who cultivate the land and plant the Indian corn, in summer; and, as soon as winter begins, they are employed in making mats, dressing skins, and in many other kinds of work,—for their first care is to supply the cabin with everything that is necessary."[33]

The new circumstance created a problem. Given traditional expectations of complementarity in gender relations, not to mention communalism and egalitarianism, the new labor burden threatened to destroy the careful balance on which the Illinois' social lives rested. But one solution to this problem was perhaps obvious, requiring just a slight shift in a key aspect of what the Illinois were already doing. To understand it we can rely on the prolific French accounts of this well-documented period, since the French could not ignore what was surely a conspicuous aspect of Illinois life. At the same time, we must be careful. The key to the Illinois' bison maximization was in the institution that French colonists so frequently disparaged, exoticized, and about which they moralized: the family. The French could not bring themselves to understand this set of relationships on its own terms, and in its full context, but we can. To understand how the Illinois preserved their community dynamics even as they maximized bison we must first consider the way all their work was organized.

◆ ◆ ◆

It is clear, if not totally explicit in the French accounts, that *families* were the central labor unit of bison hunting in the prairie ecotone. Importantly, hunters divided their scores individually, each "claiming" the animals that they personally had shot. As Râle put it, "The game is divided among the families, each receiving what its hunters have slain." In other words, while they hunted communally, they yielded separately. Crucially, they then organized the labor of processing in terms of their households. "They divide thefe Beafts according to the number of each Family; and send their Wives to slay them, and bring the Flesh to their Cabins." As Perrot put it, "Each of these families strips the hides from the animals that fall to its share."[34] It was an important dynamic: individual households effectively owned the fruits of the hunt individually.

This individualism was certainly ambiguous, and we must qualify it. After all, prairie people like the Illinois were communal in their economy. Yet what seems clear is that the communalism of bison hunting always existed in tension with a certain competitive dynamic.[35] The paradox is most visible in the culture of feasting. Male heads of households among the Illinois competed to give the most lavish feasts to their neighbors and kinsmen—working individualistically in order to maximize their apparent generosity, a key marker of status. To Perrot this aspect of prairie culture was complicated: "Although such generosity may be astonishing," he wrote, "it must be admitted that ambition is more the motive for it than is charity. One hears them boast incessantly of the agreeable manner with which they receive people into their houses, and of the gifts that they bestow on their guests."[36] Since men always competed against one another to give the best feasts and be the best providers, they had incentives to kill the most bison for their lavish banquets. This dynamic likely only increased with the initial arrival of a market for trade in bison hides. Even if the bounty of the trade was likely to be shared and redistributed throughout the communal society, rather than hoarded by individuals, the clear "ambition" for ecotone men was to be in a position to display generosity by maximizing the amount of material they could give away.

All of this led to an important consequence for individual hunters, the heads of the households that were the main unit of production in this bison economy. Not only did each man have an incentive to kill as many animals as possible, but he also therefore had an incentive to keep the most laborers in his patrilineal household in order to process the meat and the hides that "fell to his share." In other words, each man in the Illinois ecotone and in other prairie societies could benefit from having multiple women in his household. And this, in turn, is why the organization of the bison economy around individual families had an important intersection with another phenomenon: polygamy.

Although we have no idea when it began, the practice of polygamy was common in prairie society from the moment French eyewitnesses arrived in the region, and almost all eyewitnesses noted it among groups like the Meskwakis, the Ho Chunks, and especially the Illinois. La Salle noted that some men among the Illinois had as many as ten or twelve wives.[37] Given the patrilineal descent practiced by the Illinois, surely many of these men were of high status, and all the children in these families shared a common identity descended from a single father. But what seems to have determined a man's access to multiple wives was above all his abilities as a hunter: "When a man is a good hunter," Liette wrote, "it is a very easy matter for him to marry [multiple women]." As was the case in the plains cultures described by the historian Andrew Isenberg, bison hunting seems to have gone hand in hand with polygamy. Among the Illinois, individual men presided over family-based bison-producing operations in which every additional wife helped a man produce more meat, provide bigger feasts, and thus create greater reciprocal obligations from neighbors.[38]

French sources do help untangle a good deal of what was happening inside these households. First of all, polygamy was not always exploitative, as the French thought. As Liette wrote, the Illinois often practiced sororal polygyny, when a single man married several or even all the sisters of one family. In these arrangements the wives maintained great control and autonomy and lived with little discord, exercising solidarity against the husband in order to protect one another from any possible exploitation. As Perrot put it, "they always live together without any strife, all that is furnished by their husband being for the common use of their family, who cultivate the land together." Meanwhile, in these marriages, wives often relied on their own extended families—their brothers and parents—to prevent their marriages from becoming burdensome. Given that their relatives were present in the same village, wives could enforce certain expectations on the husband and resist the exploitation that the French so often imagined as the fate of Illinois women. Furthermore, a woman could leave her husband—effectively divorcing him—if disputes between her and the other wives became too intense.[39] Even as the bison economy expanded, therefore, some wives inside of polygamous marriages were protected, and complementary gender balance was preserved. For *some* wives.

If the French sensed such polygamous arrangements were less salutary for certain women, however, they were not wrong. Because while observers like Perrot noted that many women were protected from degradation because their relatives were present in the village, they also knew that some women did not have kin present. In particular, there were many women in Illinois society who entered as captives, or "servants." And even if the French were not always careful to note the distinction between these and native-born women inside of polygamous families, they also knew that these women frequently joined polygamous households as "wives." The distinction between these women and native-born Illinois women—women with extended family in the village—was crucial. A husband with a kinless wife or captive faced a much different set of expectations and many fewer constraints. He did not have to provide her family members with property and gifts during courtship, for instance. He did not have to share the fruits of his household's hunts with his wife's family for any period of time, as was sometimes arranged as a kind of dowry.[40] With a wife who was effectively kinless, moreover, the husband feared no repercussions from family members for her disaffection, nor pressure to resolve disputes and jealousy between her and other wives inside the household. Furthermore, if husbands could degrade a kinless wife, so could other wives in the family.

Here, then, is the key to an important dynamic that Frenchmen often observed among the Illinois: the strife and disharmony between the "mistress" of the household and other wives whom she considered subordinate. When women joined the household without kin present, they *were* subordinate to their fellow wives. In cases when that group of women included up to eight or ten wives, that hierarchy could become quite pronounced. Some wives could be seriously degraded. As words in the Illinois language made clear, one wife in a polygamous household was "the best loved wife" and one was "the wife who is the master of all the others." Rivalries inside a polygamous household could lead both to the isolation and degradation of one wife and to the exaltation of another: "she prevents him from going to her rival, to his second wife." These dynamics surely were at their most intense when the "second" wife in question was actually a slave.[41]

Evidence suggests this distinction between Illinois women and kinless captive women was at the heart of certain conspicuous gender dynamics Frenchmen witnessed in Illinois. The French were convinced that Illinois society was especially hard on women in general. As one French observer from this period wrote, "Perhaps no nation in the world scorns women as much as these [Indians] usually do." The French were scandalized that some women entered polygamous marriages unwillingly, since brothers at the Grand Village made marriage arrangements on their "sisters'" behalf. Father Julien Binneteau put it this way: "According to their customs, [Illinois women] are the slaves of their brothers, who compel them to marry whomsoever they choose, even men already married to another wife." Meanwhile, parents supposedly pressured their "daughters" to use their sexuality for material gain. Brothers even used their "sisters" to cover wagers "after having lost all they had of personal property." In other words, the French were convinced that "Illinois women" were generally and universally degraded. But it seems plausible that in each of these cases the French were ignoring or underestimating a key fact of these particular women's lives: the likelihood that those who were subject to this degrading treatment were not native-born Illinois women—not true "sisters" and "daughters"—but actual out-group slaves.[42]

If the French often ignored the distinction between native-born and kinless enslaved women when describing the dynamics of some polygamous families, they also did so when noting the most extreme aspect of Illinois gender relations: violence. This included mutilation, especially the cutting off of noses and ears, inflicted by "jealous" husbands on supposedly adulterous women. Liette claimed that one hundred women were scalped in Illinois during the time he lived there. As he put it, the attitude among Illinois men was that women were expendable: "There are some who say that the women are not worth the price of the least resentment." Like the lower rungs on the hierarchy inside of polygamous families, however, this violence and dismissive attitude was almost certainly reserved for women whose family could not protect them, "wives" and "sisters" and "daughters" who entered the society as captives.[43] Repeated observations of violence against women in French accounts of this period should therefore not be understood as evidence of a general degradation of all women in that society; rather, they should be read as evidence of the importance and prevalence of slavery in Illinois society at the time.

Now consider how all of these dynamics—polygamy, gender hierarchy, and captivity inside of Illinois families—impacted and were impacted by the bison economy. It is no accident that many scandalized French descriptions of Illinois gender relations come especially from the 1680s and 1690s, precisely during the bison acceleration already described. As the work of the bison economy increased, women's work became more burdensome. To a certain extent the friendly competition among sister wives, as well as women's interest in helping their households maximize production in order to demonstrate generosity and increase status, may well have provided motivation to meet new labor demands. But as those demands increased ever further in the context of the bison economy, surely the willingness of women to do this work voluntarily was strained. The ethic of gender complementarity was stretched to the limit as husbands chased what Perrot called their "ambition" of succeeding in trade. In this context women likely did resist their situation, and many would have appealed to their kin to defy their husband's efforts to increase the household production. Others appealed to priests. But some households in this context had a key advantage if they had a number of slaves present. Such households could maximize bison production by relegating to slaves the degraded and intense labor of processing. Likely a new recognition became more explicit for many households: what the bison economy called for was more wives and more women whose kin were not present in the village to defend them from exploitation. In other words, more slaves.

The evidence is circumstantial, and the French never said it outright, but as we will soon see, this acceleration of bison exploitation in the Grand Village was also coincidental with a surge in slave-raiding by Illinois war parties. As was traditional, the targets of these raids were villages to the west, where Illinois men "captured whole villages" and brought back captives. The captives, significantly, were almost all women. The men were seeking "glory," as Râle put it, but that's probably not all they were doing. They were seeking slaves, merging the old imperatives of mourning war with the new incentives of the bison economy to meet the peculiar labor demands of bison hunting at the Grand Village, even as they preserved the complementary gender relations among some members of their families.

◆ ◆ ◆

French eyewitnesses at the Grand Village noted frequently the many manifestations of military culture among the Illinois. They included the dances at which Illinois men recounted the number of prisoners they had taken, and how. There was the care with which Illinois men made up their war clubs, marking on them their totems and the tallies of prisoners taken. There was the attention Illinois men gave to their war paint and tattoos—what the Jesuits called their *mataché*—featuring symbols of their military prowess and past successes. There were the feasts that Illinois men organized to recruit comrades to a raid in order to avenge the death of a kinsman, and the ceremonies during which they entreated spirits to give them "the same speed in pursuing the enemy as [a bird] has in flying." There were the send-offs, where "everyone comes out of the cabins, dances, and gives something to these young men." There were the celebrations to welcome home war parties, and "initiation" events for the many captives that the parties brought home.[44]

If all of these were likely long-standing traditions among the Illinois, what was different in the 1680s and 1690s was the ubiquity of these aspects of Illinois culture, not to mention their scale. Warfare was a key aspect and centerpiece of life at the Grand Village. Like planting the fields, hunting, and gathering in the wetlands, there was a specific time for warfare in the annual cycle. In February, after the main bison hunt, chiefs hosted feasts, bringing dozens of soldiers together to convince them that "the time is approaching to go in search of men." Organized in both small and large parties, the Illinois launched annual attacks. Throughout the 1680s, Frenchmen frequently noted how the Illinois brought slaves up the Illinois River after their raids to the west and south, as well as from the east. La Salle noted a party returning with prisoners from the south, and Henri de Tonti observed how "not a year passes in which they do not take a number of prisoners" from the Iroquois. Setting out from Grand Village, the Illinois' slave expeditions sometimes grew massive. In 1689, two years after a highly successful raid on the Iroquois, the Illinois went west and brought back 130 captives from the Osages. In the 1680s they even placed a "habitation"—or satellite camp—on the Arkansas River, a kind of base camp for expeditions to the region. Slaves brought to the Illinois Country included Osages, Missourias, Caddoans like the Kadohadachos, and Pawnees, and, of course, Iroquois and Meskwakis.[45]

The upshot of all this was a clear ascendancy for the Illinois in the midcontinent. As Raudot explained, groups like the Pawnees, the Osages, and "other nations who are on the banks of the Missouri" were at the mercy of the Illinois in this period. "The need that these nations have for peace [with the Illinois] makes them do all that is necessary to conserve it." The Pawnees, for instance, began traveling "every year to the Ilinois to carry them the calumet, which is the symbol of peace among all the nations of the South." For the Illinois it was a sign that their power was uncontested. As Raudot puts it, "The latter are very proud to see the other nations come to seek their friendship and recognize them as their chiefs. This honor that they receive makes them believe that all the ground should tremble under them."[46]

Commenting on the Illinois soldiers in battle, numerous Frenchmen noted that the Illinois "always spare the lives of women and children." As many sources make clear, the long-dating tradition was to raid a village, kill or chase off the men, and take the women and children. In the 1690s the tradition only continued on a large scale: "They always spare the lives of women and children unless they have lost many of their own people," Liette noted.[47] By the 1680s, the largest village of the Illinois contained "many more women than men." Among Illinois men, La Salle wrote, "There is not a man who does not have several wives."[48] Conspicuously, this reflects the influx of captives in Illinois society and the hidden logic of the raids.

Evidence suggests there was clearly a new instrumental logic to the targeting of female captives in this moment, one that had a central function related to labor. As Lahontan noted explicitly about the prairie people he met during his voyage to the ecotone region, slavery and bison hunting went hand in hand. Slaves were in fact the primary hide processors during bison hunts. They carried the heavy hides to the village. Lahontan quoted his fictitious Native interlocutor by relating that hunting was a pleasant activity for the men because "slaves take all the drudgery off our hands." Indeed, if Lahontan's Indian informant can be relied upon, slaves were absolutely necessary to a successful household economy; a typical man would not even get married until he was able to accumulate a few slaves and thereby increase his ability to provide for his family.[49] In short, the massive slave raids coming out of the Grand Village during the 1680s and 1690s likely were motivated in part by this economic logic. As La Salle put it, the captives he saw in the village were traditionally the "slaves which they are accustomed to traffic," or sell to other groups in need of replacement kin. But if La Salle knew that these captives were the key to the Illinois economy and diplomacy, he also knew that they were the people "who they force to labor for them."[50]

Evidence shows that the Illinois were not the only ones turning to slaves for labor in this period, mixing the logics of mourning war with more instrumental motives when bringing home captives. For instance, in the 1670s the Iroquois themselves were seeking captives, not just to replace lost family members but also to provide household labor, or, more specifically, "because they destine them to work in their Fields."[51] But the Illinois may have done this to an extreme. Communalism at the heart of bison hunting could not "hold together" without some way to relieve the burden on Illinois women in this moment. Slaves were the answer. Words that appear in the Illinois language make it clear that captives could be degraded and treated as "dogs," like domestic animals forced to perform the will of a master. Meanwhile, slavery allowed certain native-born wives to become the *kiki8i mikintangha*, or "someone who commands other workers; master worker [overseer]." Indeed, it seems clear that many of the first wives in polygamous marriages were the supervisor of the captive women inside their households, enjoying exalted status. As Gravier put it in his dictionary of the Illinois language, the word for "the other wife"—*ni8ki8i8agana*—implies subordination and is a "term of contempt."[52] Women even held slaves personally and frequently were the ones who decided the fate of prisoners brought into the village by a returning war party. Some women evidently brought slaves with them to their new households upon marriage, both as a kind of dowry and as a key source of domestic labor by which the first wife could preserve an elevated status. In one telling case, Gravier witnessed a Kaskaskia woman praying to Jesus to make her own brother die, after he had killed the woman's slave. In other words, evidence suggests that both men and women were invested in the slave system. If each man had incentive to increase the number of "women and girls" in his household, native-born Illinois women also participated in this insourcing of degraded labor.[53] Despite the accelerating economic activity of this period, community and complementarity were preserved through slavery.

If slavery and bison hunting went hand in hand, however, the two were not linked in any strictly deterministic way. It is important to note that bison hunting did not *cause* slavery. And it was not *because of slavery* that the Illinois maximized their bison hunting. Slavery and bison hunting did not *need to* go hand in hand. But in Illinois, in the Grand Village, they did coincide. As a result, when ecological changes began to affect the bison economy at the end of the century, this was not a straightforward case of environmental determinism either. Rather, it was a rupture shaped by both human and non-human factors, including the social dynamics of slavery.

◆ ◆ ◆

Just as quickly as it reached its high-water mark, the Grand Village reached limits. The first and most fundamental change came in the form of the most obviously fickle part of the Illinois' lifeway: the bison population. Sources, of course, are not detailed enough to enable a precise accounting of the fluctuations in the bison population around the Grand Village and Pimitéoui, but evidence suggests the Illinois' large-scale harvests—especially after 1683—made an impact even on the reportedly huge herds present near the Grand Village. For instance, recall that Liette's big bison hunt took place a full eighty leagues from the Grand Village, a possible sign that either the bison herds close to the village had been depleted by human hunting or the big herds had moved away. Assuming this was true, we do have evidence to suggest *how* the bison population declined.

As we have seen, bison were a classic "commons" resource, so cultivating and preserving the bison herds in the tallgrass involved a great deal of intergroup harmony. As Hennepin once put it, groups in the ecotone had to cooperate, with a sense of the larger "publick," to ensure herd health and stability in the volatile tallgrass prairie. And yet, even as Hennepin was noting this ideal, the lifeway was under severe threat as ecotone groups went to war frequently. As Hennepin noted, "The Wild Bulls are grown somewhat scarce since the Illinois have been at War with their Neighbours, for now all Parties are continually Hunting of them." La Salle echoed this view, writing, "The buffalo are becoming scarce here since the Illinois are at war with their neighbors; both kill and hunt them continually." One underlying cause perhaps stemmed from the elimination of neutral hunting territory and certain buffer zones traditionally respected by rival groups, which became overhunted. Soon previous checks on hunting were removed and hunters killed more bison than ever, without regard for the consequences to their neighbors or to the health and stability of the herds.[54]

While overhunting was an important change in the 1690s, the declining bison population in the upper Illinois Valley may also have been connected to a larger set of environmental changes, particularly drought. Although the overall climate patterns of the centuries-long Little Ice Age had produced the cool and wet weather that probably helped to increase ecosystem productivity throughout the tallgrass region beginning around 1300 (and lasting broadly through the 1700s), the late 1600s were a clear exception. Recent research has established that the late seventeenth century in particular may have been one of the top five most significant drought events over the midcontinent during the last one thousand years. The period witnessed three major short-term drought events, including a drought measuring -4.87 on the Palmer Hydrological Drought Index in 1698. This especially dire year was surrounded by five years of extreme drought over the Illinois Valley region from 1696–99. Only three drought events over the midcontinent were more severe during the previous millennium, including the Dust Bowl and some sixteenth-century megadroughts.[55]

It is difficult to be precise about how drought may have impacted the local ecology on the ground. But, specifically in 1698 (the most severe year of the drought), the priest Jean de St. Cosme traveled through the Illinois Valley and noted a desiccated landscape. One of his fellow travelers noted that the Illinois River was extremely low—"no water in the Illinois," as he put it.[56] As the party moved past the locations of the Grand Village and Pimitéoui, moreover, St. Cosme noted curious facts about the bison. On the one hand, he reported that they were still present nearly everywhere. On November 11 members of St. Cosme's group killed four bison, marveling at the ease with which the animals could be taken. Yet there was something wrong: they were all extremely skinny. "They were so thin during this season that [the hunters] only took the tongues."[57]

To understand the significance of this report we must recall certain facts about the bison themselves. According to biologist Dale Lott, the bison's annual cycle centers around a "fat economy": the animals fill their fat stores through the summer months, as the massively productive grassland translates the sun's energy into calories and the animal's rumen transforms those calories into huge bodies, or "reservoirs of biomass."[58] In a typical year those bodies are at their *biggest* in the fall. Perhaps this was even especially true in the rich grasses of the tallgrass prairies of the period. As Hennepin wrote in 1680, "the Flesh of these Beasts is very relishing, and full of Juice, especially in *Autumn;* for having grazed all the Summer long in those vast Meadows, where the Herbs are as high as they, they are then very fat."[59] Something was different—very different—in 1698. St. Cosme's is the only European report from Illinois Country ever to mention malnourished animals. Especially for hunters trying to capitalize on the large herds of the region, that is, expecting and needing the animals to be "full of juice," the scraggly, lethargic animals instead probably represented a crisis. Malnourishment almost certainly reduced the size, not just of animals but of the herds themselves.

There were likely other subtle impacts caused by this climate event. In particular, while the Illinois exploited bison to an extraordinary degree in this generation, they also relied on other aspects of the ecotone environment for their agriculture and especially their gathering. It was the diversity of the region, after all, that made their lifeway—the mix of resources from wetlands, lowland forest, upland forest, prairie, and aquatic environment alike—so prosperous. Most important in many ways was their use of the forests that hugged the river valleys and extended into the uplands, where La Salle noted abundant "forests of high trees" interspersed in the prairies. As ethnobotanist Michael Gonella has shown, this component of the landscape—in the Miami-Illinois language called the *mihtekwaahki/mihtehki/ahtawaanahki*—was always the most important limiting factor of prairie peoples' lifeways. Though forests could be extensive, the resources in them were widely dispersed and prairie people like the Illinois needed a relatively large amount of forest to satisfy their needs.[60]

Climate conditions in the late 1690s may well have posed a problem in this connection. Evidence is scarce, but it seems like a timber shortage was already a factor for the Illinois as early as 1693, when many residents of the Grand Village moved a day's journey to the south, to the new village Pimitéoui. They did this, La Salle wrote, because they had used up the nearby timber at the Grand Village and because "their firewood was so remote."[61] Arriving in the new location of modern-day Peoria, they hoped for more robust timber resources and hardy forests. According to linguist Michael McCafferty, the name "Pimitéoui" translates to "fire passes through" and may even refer to the local forest's resilience against fire, reflecting the importance of forest cover in the logic of the village's location.[62] In the context of the droughts of the late 1690s, however, it is easy to imagine forest quality in the entire region—including at Pimitéoui—suffering, especially here on the blunt edge of the forest-prairie tension zone. If the ecotone in general represents a kind of "battle zone of floristic groups," as in the words of John Madson, a drought cycle that lasted from 1661 through the turn of the century and climaxed in the 1690s may have impacted the resilience of the forest and caused the Illinois to seek forest resources elsewhere. In short, drought likely affected the Illinois' strategy well beyond its impact on bison numbers.

A final biophysical event impacted human history at Grand Village in this period: disease. As we will see, European diseases had been impacting the ecotone's large villages since the 1660s at least, and possibly before then. Finding hospitable "disease ecology" within concentrated populations, where people were constantly coming and going, diseases became more virulent than they might have been in the protohistoric period, when the earliest European pathogens began circulating around North America. Given the influx of foreign people—in the form of slaves, missionaries, hide traders, and constant visitors—it is no surprise that the Grand Village witnessed the first recorded epidemic in Illinois history around 1694, when Jesuit Jacques Gravier noted many in the village were sick. There is no way to measure the scale of the outbreak nor its demographic costs. What is clear is that different responses to the pestilence created strife among different community members, as the Peorias not unreasonably blamed Gravier for the unusual illness, while the Kaskaskias accepted the priests' spiritual power as a possible response to the epidemic.[63] Equally significant as a challenge to the intragroup harmony in the Grand Village as a demographic event, the outbreak of 1694 was an important change in the Grand Village's reality, and it foreshadowed future dangers of the Illinois' bison-based economy.

In any event, nonhuman factors all contributed, and the Grand Village started to come apart in the final years of the seventeenth century. After working together to help shape the great acceleration of bison hunting, slave-raiding, and community formation in this unprecedented population center over a generation, the nonhuman agents of local ecology helped accelerate the demise of this special assemblage. Environmental changes impacted Grand Village's economy, not to mention how they affected the community dynamics that were at the center of the Illinois' massive population and maximizing strategies.

◆ ◆ ◆

Even as the Grand Village began to come apart in a physical sense, however, the Indigenous motives underlaying it continued. In 1699 the Kaskaskias, the largest ethnic group within the loose collective of Illinois-speaking peoples, relocated their entire lives once again, moving from the relatively new village at Pimitéoui to a completely new location on the Mississippi River near modern-day St. Louis. A dramatic end to the great consolidation at the Grand Village, this radical move was also an effort to continue the enterprising activities that had defined the great acceleration of the previous generation.

Indeed, the Kaskaskias' migration was the most significant and poorly understood move in the history of the ecotone, a continuation of their attempt to take advantage of key strategic advantages in their dynamic region. In 1699 La Salle's dream of a colony in Louisiana became a fledgling reality when Pierre Le Moyne d'Iberville finally began the official colonial project that would become Louisiana at La Mobile. The Kaskaskias surely knew that one of the earliest economic enterprises planned for the Louisiana settlement would be an ambitious attempt to fulfill La Salle's long-held dream for a bison tannery. Often ignored by historians, this project was a key event. Abandoning a changed and degraded upper Illinois Valley location, where they had thrived uneasily for a generation, they were by no means leaving behind the violence that was entangled with their recent enterprise. Indeed, that side effect of the great bison acceleration was poised to continue.

6

HIDING IN THE TALLGRASS

TRAVELING THROUGH THE ILLINOIS VALLEY IN 1711, JESUIT GABRIEL Marest witnessed an Indigenous world torn by violence. Recalling how over the course of that summer his small party had made the long round trip up the Illinois Valley to Michilimackinac and back, Marest reported that the region was full of "hostile bands" on revenge attacks, "parties of warriors who range the forests." Battles and raids were ubiquitous. Indeed, the risks of traveling in the country were so great that few people undertook it; starting from the new settlements at the Kaskaskia River on the Mississippi River, past older Illinois settlements at the former Grand Village and at Pimitéoui, Marest and his Indigenous guides sensed that the country had become a no-man's land, a nearly trackless wilderness with barely "any path which could guide us." As Marest recalled, his group "traveled for twelve days without meeting a single soul" save for ominous raiding parties. Animals were abundant but hunting was dangerous; hunters in their small party feared "being discovered by the sound of the shots." Blazing a new trail and petrified of being discovered by war parties that scouted everywhere in the hazardous region, Marest's party moved slowly and deliberately, keeping "on our guard . . . in order not to meet them."[1]

The violence that Marest described in the Illinois Valley in 1711 is notable for its seeming intensity and pervasiveness. But it is equally notable for its timing. In 1701, after years of working to create a stable alliance system of

Iroquois and Algonquian peoples against English power in North America, administrators in New France had celebrated what became known as "the Great Peace," a time when hundreds of Indigenous leaders traveled to Montreal to sign their names to a covenant with "Onontio," the *Great Father*, Governor Louis-Hector de Callière.[2] In signing the 1701 Great Peace, representatives of nearly every Indigenous group of the Great Lakes pledged to end the Beaver Wars and henceforth mediate their disputes peacefully, promising to avoid the very kind of violence that in 1711 seemed to engulf the Illinois Valley. Whether because other signatories were more intentional in following the promises of the 1701 treaty or for some other reason, no other place in the Great Lakes or Mississippi Valley during this period seems to have witnessed such severe conflict in its wake as the ecotone, especially the Illinois Valley. So how do we understand where that violence came from?

Surely part of the answer lies in the regional history we have already explored: the long processes of migration and encounter by which foreign groups like the Illinois and the Meskwakis and Sioux all came together in the tallgrass borderlands beginning in the early 1600s. But there was something new going on there as well, or at least a new source of rivalry. In 1699, just as the plans for the diplomacy at Montreal were picking up, another French colonial project was getting underway at Louisiana. But while the new colony of Louisiana would eventually focus on agricultural development, mostly in the lower Mississippi Valley, one of that colony's first enterprises—and a central *raison* for the colony, dating back to La Salle's early colonial scheming in the 1680s—constituted an immediately transformative event for the Indigenous residents of the mid-Mississippi Valley. In 1702 a Canadian royal judge and fur trader named Charles Juchereau de Saint-Denys brought a considerable party of French colonists, a large amount of material, and an equally large set of ambitions to the prairies adjacent to the Wabash River Valley in Illinois Country, where they aimed to establish a special and unprecedented kind of colonial outpost: a *tannery*.[3] Not just any tannery, moreover, but a tannery for the hides of what they called the "boeuf sauvage," and even the "boeuf Illinois." To us: the bison.

Juchereau's aim was not modest. As he saw it, in terms of value the bison hides would soon outstrip beaver furs, deerskins, and other pelts from North America, and he predicted that before long he would easily be exporting up to one hundred thousand hides per year. The prairie edge of the Wabash Valley was one of the ecotone's greatest bison territories, and the tannery attracted Illinois-speakers and the Mascoutens who moved in to concentrate their communities—and their labor—for the purpose of supplying the new tannery project. In a matter of two years they produced an extraordinary fifteen thousand hides.[4] Yet while Juchereau and his Louisiana partners may have wanted to eventually staff their tannery with French artisans and enslaved Africans working along the model of a European factory, it is obvious that almost all the work to create this initial output was to be performed by Indians. Here is why the tannery was such an important event, for it interacted with and amplified ongoing trajectories. Given how the prairie Indians' labor system so importantly entangled with captive raids and the imperatives of mourning war, the tannery only increased the demand for captives and thus intergroup tensions, even as it attracted the Illinois to a location closer to the source of captives in the west. In short, the hide trade accelerated the slave-based violence of the ecotone to an unprecedented scale.

Meanwhile, the tannery also accelerated something else: disease. The tanning enterprise relied on crowded hide-processing villages, filled so importantly with Frenchmen and captive laborers from faraway villages. Not only did these settlements create a ripe ecology for the transmission of disease, but their foreign captive laborers likely became the vectors for the introduction of a specific scourge: smallpox. Two major outbreaks in a Mascouten and a Kaskaskia village decimated these large villages. Entangling with local circumstances to produce massive effects, diseases now created new disruption among the Illinois and Mascoutens at the tannery. They also importantly infected Frenchmen, including Juchereau himself, who died in a Kaskaskia village in 1704.

Juchereau's quick death has led historians to minimize the historical significance of the tannery as merely a flash in the pan. But viewed in another way, the explosive and temporary nature of his intense project is precisely what made it so momentous to the Indigenous people of the eastern prairie region. When Juchereau's tannery fizzled, the Illinois quickly lost all the benefits that were supposed to flow from the risks and sacrifices of their lifeway and recent migration. Moreover, the end of the tannery in 1704 was not just a short pause or disruption. For the Illinois in particular it may have seemed potentially permanent. The reason is simple. Although Frenchmen viewed the Wabash River Valley of late- seventeenth century as the most prolific bison habitat they had ever seen, they were of course naive. By the early 1700s their new colony at Louisiana allowed the French to conduct more extensive expeditions up the Missouri River, expeditions that introduced them to an important fact: the Wabash River Valley was by no means the most extensive bison habitat of North America! Thus it was that when French survivors of Juchereau's initial project began to restart the bison hide trade in the Mississippi Valley—something they did only haltingly in the years after Juchereau's death—they did so slowly, and only partially in the Wabash Valley, where the Illinois and others waited expectantly in the tallgrass. Instead, they more frequently went to the Missouri River to try their bison project on the plains, bringing the mixed benefits of colonial trade to people who were actually rivals and even open enemies of the Illinois: the Osages, the Padoukés, the Pawnees.

Here was a familiar dynamic for the Illinois. Just as they had long felt at a disadvantage in the beaver trade, owing to their position at the southern edge of prime beaver habitat, they now understood that their position in the eastern prairies was—in comparison to the shortgrass habitat of the Great Plains—relatively marginal bison habitat. The *edge effect* turned against them. They were left with the grim consequences of maximization. The much-anticipated hide trade barely limped along. What remained were the intergroup animosity; the demographic losses and motives for new mourning war; and the polygamous households full of captives generating hostilities with neighbors. By the 1710s, as Marest noted, resentful enemies raided the Illinois Valley, seeking their revenge on the Illinois speakers and looking for replacement kin. Importantly these sought-after kin included the Sioux, the regional hegemons with whom the Illinois had long tried to forge peace. The Great Peace was no match for the Indigenous conflicts brewing inside the ecotone, on Indigenous ground.

◆ ◆ ◆

In 1699 the Kaskaskias, the largest ethnic group of Illinois speakers, arrived in the American Bottom, having abandoned the massive population centers in which they had concentrated for two generations in the upper Illinois Valley. By moving they had left behind their Peoria kinsmen to join several other groups, particularly the Cahokias and the Tamaroas, who lived in a village on the east bank of the Mississippi River not far from its junction of the Missouri. Occupying this region, they were located at a strategic place where middle America's major rivers—the Mississippi, the Missouri, the Illinois, and the Ohio—all come together. The Kaskaskias took up residence in the heart of the Bottomlands, near the place where the Mississippi is joined by the river that carries their name.

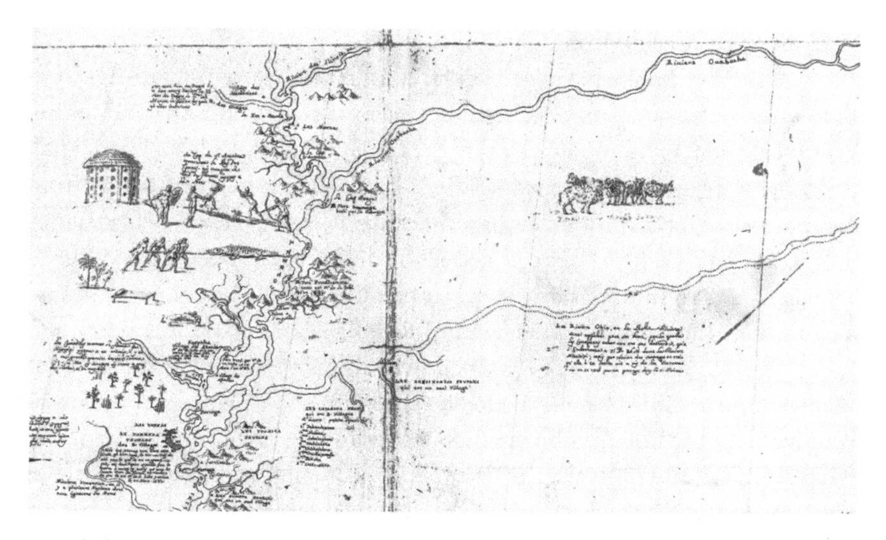

Untitled manuscript map of Mississippi Valley (detail). From Vincenzo Coronelli, 1683[?]. Coronelli placed this wonderful illustration of "Pisikious, or boeuf sauvage" in the Wabash Valley, which French observers understood to be the most prolific bison habitat in North America in the late 1600s. Image courtesy Beinecke Rare Book and Manuscript Library, Yale University.

A telling perspective of the riches of their new home comes from a French traveler called André Pénicaut, who traveled there from lower Louisiana several times in the early 1700s. Here in close proximity were river valleys teeming with wetlands and abundant bottomlands forests, as well as the heart of the "western forest prairie"—an upland mosaic rich with game. As he passed through Pénicaut took special note of the abundant grasslands, commenting especially on the "delightful prairies" and the "very pretty vistas" near the Illinois' new village at Kaskaskia. They were full of animals. The bison were especially abundant on the Ohio and its tributary, the Wabash River, or, in the Miami-Illinois language, Waapaahšiki Siipiiwi.[5] Sparsely populated, the prairies adjacent to the Wabash had long been Myaamia hunting territory, a vast "wilderness" [desert] extending "one hundred and twenty leagues [240+ miles]" and teeming with bison.[6] A notation on the famous map of the midcontinent created by Guillaume de L'isle in 1718 shows the prairies adjacent to the Wabash as prime bison-hunting territory.

In moving to the region, the Illinois were seeking opportunity; after the changes of the late-seventeenth century, this landscape featured several advantages over the upper Illinois Valley. But the reason for their migration was not about the environment alone, for indeed they were recruited. The Kaskaskia chief Rouensa arrived in the confluence region in 1700, boasting of his ties to the newly established colonial project in Louisiana and telling everybody that he was "called by the great chief of the French [in Louisiana], as Father Maretz [Marest] has told him." To be sure, a separate report from Jesuit Jacques Gravier from a few months earlier calls into question how enthusiastic Marest might have been about the Kaskaskias' new connection to Louisiana and its future. In 1700 Gravier preferred that the Kaskaskias should stay in the Illinois Valley, at least until the Louisiana plan was clearer and more specific. As he put it, he wanted the Illinois to wait to find out "what place the great Chief who is at the lower end of the River wished them to remove."[7] But if Rouensa boasted about it, and Gravier worried about it, what seems abundantly clear is that the Illinois' migration was partially in response to a surprisingly elaborate plan hatched at the start of the Louisiana colony. For the French, attracting the Illinois to the Wabash was part of a long-standing vision dating back a generation to capitalize on the region's bison. Celebrating the prospects, Pierre Iberville wrote that the prospect of a new hide trade in the Wabash was "of greater consequence than can be believed."[8]

However grand and seemingly fantastical, it is important to note that Iberville's plan for bison exploitation already had a long history. In fact, there is no doubt that the bison hide trade was La Salle's original and primary idea for commerce in Louisiana. As discussed earlier, La Salle did in fact open a small bison trade with the Illinois at Fort Saint Louis, and he told his subordinates there to encourage the Illinois "to hunt as many bison as they could." This reflected a larger plan that he hoped to fully realize in time. Reporting to Colbert about his plans in the 1680s, he told the minister that "the mere trade in the hides and wool of the wild cattle which the Spaniards call 'Cibola' was capable of building up a great commerce and supporting powerful colonies." Dramatically, La Salle had shown Colbert a sample bison hide gathered during his initial exploration in the Illinois Country. Using this to demonstrate the difficult logistics involved in transporting massive bison hides upriver to Canada by canoe, La Salle thereby justified his new colonial project centered on the Gulf of Mexico to which large quantities of bison hides could be effortlessly floated downriver on big boats. Here was the essence of the project that Colbert actually approved and the centerpiece of his grant: "the sole privilege of trading in the skins of the wild cattle."[9]

Bison hides were always at the center not just of La Salle's grants, but of La Salle's ambition. Perhaps because the plan proved so quixotic, La Salle's writing on bison is often ignored. But it is extensive. *Listen* to his enthusiasm about the Mississippi River: "All the country that is along this river . . . is full of wild cows, more than can be said." Of the Wabash in particular he wrote, "The abundance of oxen is greater than one could believe. I saw a single band of Indians kill twelve hundred in eight days." The economic opportunity was great, La Salle wrote, because the Indians of the prairie region did not know the value of the hides: "We can have their leathers and their skins at a low price, these Indians not yet knowing our commodities." And if the Indians did not know the value of the skins, meanwhile neither did the French. "It will be up to me to put a price" on the bison skins, La Salle wrote. "We must not doubt that we should expect a great profit." Early estimates given in La Salle's writings suggest that they could sell the hides at wholesale markets in France or in the Caribbean for something like 24 livres each. If this had proved true, then the value of the hide trade as anticipated by La Salle would have easily eclipsed the entire value of the beaver trade.[10]

Meanwhile, logistical challenges involved in this enterprise are probably why La Salle also began scheming about something else: a tannery. He planned to put a tannery in the middle of bison country—on the Ohio, in reach of the Wabash River Valley—and he would then seek permission from the king to trade the output of this industrial installation without restrictions in France; he wanted "free entry for tanned leathers, as [if] from the Kingdom." In time workers could begin *raising* bison for their hides, for bison could be tamed by separating them from their mothers at a young age. Meanwhile, Indians would continue to hunt "a great number" of wild animals. African slaves could be imported to staff the tannery, and "one could teach them to be the craftsmen" at the leatherworks. "There," wrote La Salle, "is everything you need." It was La Salle who originally speculated that with time French hunters on horseback could bring the annual yield of animals to upward of one hundred thousand animals. In the early years, even with no additional help, the pedestrian hunters—the "Miamis and Islinois alone"—could be counted on to produce ten thousand per year.[11]

When La Salle died in 1686, Iberville simply inherited his vision. Writing of prospects for the new colony after arriving at Mobile, Iberville once again sent sample bison hides for French officials to examine. As he put it, "I have eight skins of wild bulls and cows and their wool to serve as proof" of this new commodity's value. He also sent balls of bison wool "spun by the Indian women and one blanket made by them to show how this wool can be worked like that of sheep." Profits would not be too far off: "I hired the travelers who went back to Illinois to hunt bison, [promising] we would pay seven livres and even more per hide, if it was found to be good in France, and to [explain the plan to] all the Indians, who much prefer to hunt bison than beaver." As Iberville contended, the only obstacle "in the beginning" would be the difficulty Indians faced in transporting the huge hides. But soon, "the Frenchmen there could build barges and flat boats, to deliver *thousands* at a time" (emphasis added).[12]

So it was that in this context an unlikely project got off the ground in the very earliest years of colonial Louisiana. With the Illinois now in place, the little-known colonial agent Juchereau actually aimed to make all the talk about the bison hide trade real. The "tannery" would not be just a dream; with Juchereau and his soon-to-be Indigenous partners it would be a messy, physical, bloody fact. In becoming reality, however, it would quickly become quite different from how he and his predecessors imagined it.

◆ ◆ ◆

In 1702 Charles Juchereau de Beaumarchais de Saint-Denys was well qualified to lead a new enterprise in the North American fur trade. Born in 1655 in Quebec, he was the son of Nicholas Juchereau, who had been a member of Canada's Company of One Hundred Associates, the proprietary group of investors holding a monopoly interest in the fur trade and economic development of New France from 1627 until Canada became a royal colony in 1663. Charles Juchereau had grown up in contact with the fur trade and, after acquiring considerable wealth in the form of a dowry from the family of his wife, Thérèse-Denise Migeon, invested enthusiastically in fur operations. But Juchereau began his career at an unfortunate time; the market for beaver furs in the Great Lakes in the late 1600s was not only heavily controlled by a jealously guarded monopoly, it was also glutted. And so in the late 1690s Juchereau traveled twice to France with a secret agenda to make a case for an alternative project. Through his marriage he had become the uncle to Iberville, and this connection surely gave him an interest in the new colony at Louisiana. While in Paris he proposed to the king a new scheme, enlisting the support of an ally at court, the Comtesse de Saint-Pierre. Clearly it took all of his connections to win his new privilege, but he soon had his permission from the king and a monopoly for Louisiana.[13]

News of Juchereau's concession was published June 4, 1701, just weeks before the Great Peace was to be signed in Montreal. The heart of the enterprise would be La Salle's idea of a tannery. Returning to Montreal in 1701, Juchereau not only began recruiting the full complement of hired men and *engagés* stipulated by the terms of his grant, but he secured a massive amount of credit in order to purchase the material, loaded into eight boats, on which this serious enterprise would depend. He formed a new company, the "society of the Mississippi," in order to attract investors. Evidence suggests he took on 40,000 livres of debt, an amazing figure. But recall that La Salle estimated 10–14 livres of profit from every bison hide exported to France. If La Salle's estimate of one hundred thousand hides annually proved even close to correct, the tannery would be an enormously profitable venture.[14]

What was a tannery? It was an industrial solution to a basic biological problem. Given that a raw animal skin rots if left untreated, tanning makes it impervious to the decomposition that naturally occurs on dead skin. The process is accomplished with a series of chemical treatments to remove and neutralize bacteria that cause putrefaction and to strengthen the natural fiber construction of a hide into the strong flexible material known as leather. An ancient practice, tanning was transformed in the eleventh century into a relatively regularized craft in Europe. In the colonial context, tanning quickly became foundational. Although animal skins and tanning materials were abundant in many colonial contexts, leather could not make itself. And nothing else could happen, observers like Alexander Hamilton wrote, if the workers—including, importantly, slaves—did not have shoes. Leather tanning thus was among the first and most important industry everywhere colonists went, first to supply local need and then as a prospect for export, either to metropolitan markets or to other colonies and especially to plantations.[15] But tanning is a complex activity and was one of the most complicated and demanding trades in the premodern period—demanding material, infrastructure, and, most of all, labor.

There were two main means of tanning in the premodern period. Indigenous hide tanning most frequently utilized a concentration of animal fat, usually created primarily from a concoction of brains and other organs, along with a process of smoking to help the oil penetrate the hide, kill bacteria, and soften fibers. Europeans also performed oil tanning, for instance, to make chamois. But the more common kind of European hide tanning was vegetable tanning. It derived tannic acid most frequently from the bark of oak or hemlock trees; hides were soaked in a tannic liquor made from the bark to kill off the bacteria within the hide and arrest the putrefaction of the skin. Starting with a raw or salted hide, the tanner put it in a bath with lime or a fermented barley solution to commence "limning" or "raising," a process to make hair follicles easier to remove. Once softened and "raised" for several weeks, the hides were then "beamed," or stretched across a frame and scraped free of hair and imperfections. Then the real "tanning" commenced, when the hide was soaked for several *months* in the tannic acid. The mill used to grind the bark to make this acidic liquor thus was a central part of the tannery's machinery, while the soaking vats were the biggest part of the operation's footprint and a typical bottleneck of production. When the soaking was done, workers dried the tanned leather and squeezed out all the moisture. Like beaming, this was a particularly labor-intensive part of the process. After all these steps were finished, the resulting material could be treated with oils and other materials by a currier to produce a finished leather. Start to finish, one French military manual from the 1790s estimated, the process could take six to nine months or more.[16]

Tanning could be done on a small scale, but a tannery was an application of the process on an industrial scale. Some eighteenth-century economists agreed that leather-making was the most important preindustrial craft; the value of the leather industry in England was second only to wool, and in France it notably exceeded the value of metal. Given leather's role as an input to manufacturing as well as its centrality to military equipment (particularly the outfitting of cavalry), many perceived a shortage of leather production to be one of the great threats to strategic power in early modern Europe.[17] A good deal of energy thus was spent studying the industry, protecting it, and growing it. The resulting literature gives us a good understanding of the materiality of tanning in fairly precise detail. Illustrations from Henri-Marie Duhamel du Monceau's *Descriptions des arts et métiers* (1761–88) as well as commentary in various trade manuals and economic treatises—not to mention Diderot's and d'Alembert's *Encyclopédie*—help us understand the elaborate infrastructure of a typical eighteenth-century French tannery. If we think about Juchereau's tannery in 1702 as both a planned environment and a lived reality, we get some clues as to what the French planners may have been *trying* to realize on the ground.

Juchereau was not the first to try to establish a tannery in France's North American colonies. The earliest large-scale tannery in Canada was founded in the 1660s. Aiming to increase the economic self-sufficiency as well as the export of the colony, Intendant Talon sponsored a merchant named François Bissot with some 3,000 livres to start a tannery near Quèbec. Bissot worked together with a newcomer from Poitiers, Étienne Charest, who was familiar

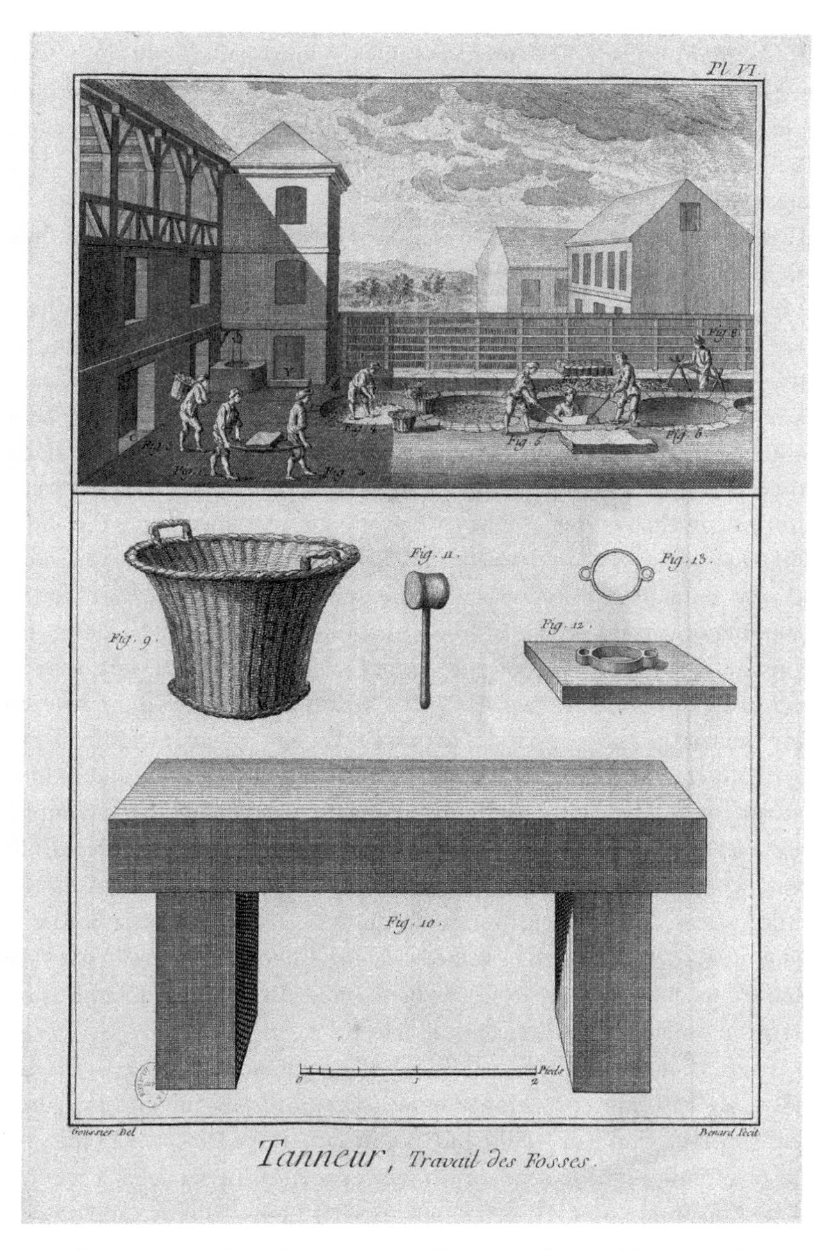

Plate depicting an eighteenth-century French tannery: "pit work." From Diderot, *Encyclopédie* (1751–66). Image courtesy Bibliothèque nationale de France.

with the tanning trade. The enterprise quickly flourished. Importantly, Charest's tannery worked not only the hides of domesticated cattle, but also the elk and deerskins hunted in the interior, presumably by Indigenous people. When Charest died in 1699 he was one of the richest colonists in Canada. It seems possible that Juchereau knew Charest; he probably knew how successful his business was. Interestingly, Charest primarily served only local shoemakers and beltmakers; he never really developed an export market.[18]

Export was surely the bigger prize for an enterprising colonial businessman. In fact, Talon's original plan for Charest's tannery was to produce leather for export to fulfill a specific market demand. In 1667 Talon had asked Bissot to conduct an initial experiment: use moose skins to make *buffle,* or "buffalo." This term "buffle" referred not so much to a specific animal, but rather to a specific kind of leather: a big thick hide that was used especially in outfitting the military.[19] They were often made from oxen, but Talon thought moose or elk could substitute. The market was potentially very large. Military manuals in France detailed the importance of buffle hides for the equipping of various classes of troops, including, especially, cavaliers, dragoons, and cannoners. For instance, each French cavalryman wore a special "vest of buffalo skin, named *le buffle.*" Meanwhile, the military required these same large and tough hides to make other equipment, including straps (especially the shoulder strap known as the *bandoulière*), belts, and saddles. Overall, one study in the late eighteenth century estimated that a complete army of 105,000 troops would require annually 15,909 buffles. Naturally, these would all be "imported from abroad," as there were no animals in France whose skin was appropriate to the task. Talon may have been conscious of this demand when he helped establish Charest's tannery in the 1670s; he challenged Bissot to try to supply up to fifteen thousand moose hides as a reserve of buffles for the French military.[20]

If he was aware of this demand for buffle, however, Talon may well have known about a tanner in France named Everhard Jabach, who was also a member of the Company of the Indies and a financier. Precisely in the same year that Talon began to sponsor Charest's endeavor (1667), Jabach acquired permission from Louis XIV's famous minister Louis Colbert and Secretary of War François Michel Le Tellier to establish a tannery for le buffle in Corbeil (now Corbeil-Essonnes), outside of Paris. Jabach's buffalo hide factory was built explicitly to meet military demand and it quickly grew, importing hides from different parts of Europe (especially Hungary) and tanning them

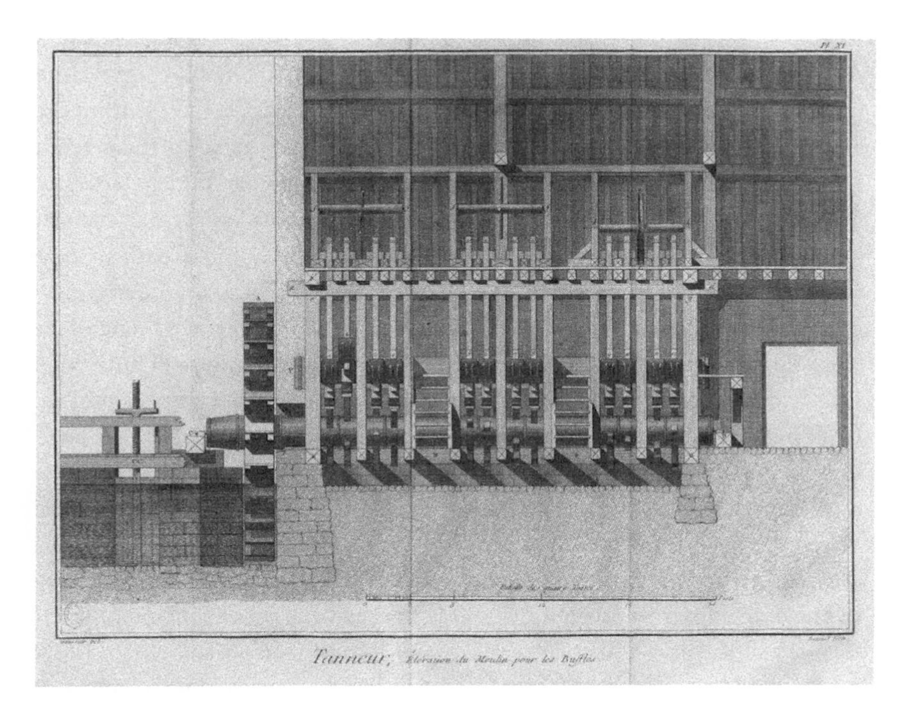

Plate illustrating the fulling mill machinery at Jabach's tannery for buffalo hides in Corbeil, France (now Corbeil-Essonnes, near Paris). From Diderot, *Encyclopédie* (1751–66). Image courtesy Bibliothèque nationale de France.

to the specifications of le buffle. The output was massive, amounting to an estimated six hundred hides per week. Needing workers to staff the massive operation, Jabach found orphans and foreign workers to train in the tanning process.[21]

By the late eighteenth century the tannery at Corbeil was such an elaborate production facility that it became famous in France.[22] Illustrating the *Encyclopédie,* Diderot's engravers used the Corbeil factory itself as a model for a tannery, depicting its complex mill equipment. Jabach himself was considered "the most considerable tanner of his time" and with his wealth became a major patron of the arts. The massive scale of his Corbeil operation reflects the size of the market in these specific hides, the exclusive product of this tannery.

Whether Talon, Charest, or La Salle actually knew about Jabach or the Corbeil tannery as they each schemed to establish a buffle tannery in North

America is not clear. We simply do not know if Talon's plan to export moose hides from his Quebec tannery or La Salle's early dreams of expanding the bison hide trade at Illinois—or his request for "free entry" of hides from Louisiana to the French market—were attempts to compete for Jabach's contract with the military. Whether Talon, Charest, and La Salle did or did not know about Jabach's enterprise, however, it seems almost certain that Juchereau did, and that he viewed the North American bison as a likely new source for the special buffle hides that were so in demand by the French cavalry. And so the model of industrial tanning at Corbeil may well have been what Juchereau eventually wanted to replicate in his Wabash tannery project. Understanding the Corbeil factory's precedent, in other words, helps us to imagine what might have been on Juchereau's mind—and in his boats—as he and his men loaded up the company's large investment of equipment for the new project. Did they plan to construct a mill for grinding bark? Did they envision a massive tanning yard with vats for soaking hides? Did they intend to build machines to speed up the labor-intensive processes—beaming, for instance, or fulling? This may well have been Juchereau's vision.

Plans aside, the reality was something different. However much Juchereau may have wanted to replicate Jabach's tannery in Corbeil, or even Charest's more modest tannery in Quebec, circumstances on the ground were obviously challenging in the interior of the Wabash Valley. Any ambition to establish industrial infrastructure in the midcontinent was clearly a long-term vision. In the early years, even with the eight boats and 40,000 livres of investment that Juchereau prepared to bring to the Illinois Country, the setup at the Wabash would by necessity be something much more humble and rudimentary. Yet here is the important fact: Juchereau's project did in fact become hugely productive, and immediately. By 1704, the year he died, Juchereau's project produced fifteen thousand hides, fully a quarter of the number of hides that the massive factory at Corbeil had produced in the same period and more than the heady estimate that La Salle proposed in his boosterish reports. It is an astonishing number. How did they do that? If it wasn't achieved by replicating the techniques and technology of Corbeil's factory in the backcountry of America (which is clear), and if it wasn't accomplished by importing the labor system of the guild or the plantation and applying it to bison tanning (also highly unlikely), then something else was happening. It is not difficult to imagine what that something was: the local Indigenous people. To produce fifteen thousand hides in a single year, the newly arrived and relocated Indigenous populations of the Wabash valley burst into furious activity, just as La Salle predicted they would, making *Erenans8ie*, or bison hides.[23] And although their work *process* looked nothing like that of a tannery in a town like Corbeil, the scale of the work was close.

◆ ◆ ◆

Juchereau left Montreal with his party on May 18, 1702, and entered a region roiling with hostilities. In spite of whatever promises they had made at the Great Peace in 1701, groups present in the ecotone remained antagonistic. In particular, rivalries among the Sioux, the Mascoutens, the Meskwakis, and the Illinois seethed as they competed for access to traders from Canada and the new Louisiana colony. For years the Sioux exercised dominance by recruiting French traders to work with them, and they succeeded especially in their relationship with Pierre Le Sueur, a *coureur de bois*–turned–colonial agent who in the 1690s established a new route to Sioux country from Michilimackinac. For their part, the Foxes and the Mascoutens opposed Le Sueur's activities and what they feared as Sioux hegemony. To keep the peace and appease the Meskwakis and other Algonquians, Canada had recalled traders from the Sioux, including Le Sueur, even after he established a trading post at Chequamegon Bay in 1693. Undaunted, Le Sueur then joined the Louisiana effort and set up a new outpost among the Sioux under Louisiana authority, called Fort Huillier. Located on the Rivière Verte, or Blue Earth River, this is where Pénicaut and his companions killed four hundred bison and joined in the project of a new trade enterprise on the northern edge of the prairie peninsula.[24]

Although not sponsored by Canada, Fort Huillier was no less distressing to the Foxes and the Mascoutens, who attacked Le Sueur's outpost in the summer of 1702. When Juchereau entered the Mississippi Valley by way of the Fox-Wisconsin portage in the late summer of 1702, intending to travel to the Illinois Country, the Meskwakis and the Mascoutens intercepted him, suspecting correctly that he was planning to establish a trading outpost among their other rivals. They demanded a toll of "a thousand crowns" in exchange for allowing him to continue to his destination with his men.[25] Herein may have been an important moment in the positioning of Juchereau's project within the Indigenous politics of the ecotone. It seems likely that Juchereau would have assured his Meskwaki and Mascouten hosts that his project would counterbalance the trading opportunities that Le Sueur had brought to the Sioux. He even may have tried to recruit both groups to assist in the new project, just as Iberville had previously recruited the Illinois. In any event, Juchereau's new project clearly entered a context of competition and rivalry among ecotone groups. Importantly, as Juchereau continued south, bound for the Kaskaskia village in the fall of 1702, the Mascoutens followed him.

Juchereau's arrival at Kaskaskia must have been a dramatic moment. It was the first meeting between himself and the Illinois speakers, coming only a few months after the Kaskaskias' relocation from the Illinois River villages. We can only imagine the scene as Juchereau pulled his boats to shore at the village, laden with material and twenty-four men (the largest French party every gathered this far west in Illinois Country. To Jesuit Gabriel Marest, Juchereau's arrival was auspicious. As he wrote after observing the meeting of the Kaskaskias and the tannery's new architect that Juchereau was "prodigal of his promises, but thinks, in reality, of his interest." To be sure, his Indian partners—Illinois and Mascoutens—thought of their own interests too. The Illinois even acted standoffish, refusing to relocate their village too close to Juchereau's chosen site.[26]

What happened next? One way to understand what unfolded as the "tannery" took off is to keep our eye on the material, that is, what those eight boats contained. We know his men quickly built a fort about two leagues from the mouth of the Ohio River. Before stashing the cargo inside the fort, Juchereau probably began to reveal the boats' contents to the Kaskaskias and the Mascoutens, and presumably others. Surely some of the cargo comprised essential material for building a factory. But a bigger part was something else: trade goods. Bring us bison skins, Juchereau must have told his new partners, and here is what we will give you in return. The conversation likely went back and forth. As La Salle had said a generation earlier, no one knew exactly what bison skins might be worth, so setting the price was a complex negotiation. Evidently Juchereau offered an especially valuable commodity—sixty cups of gunpowder—as a price for a single bison skin.[27] The timing was perfect, as the Illinois and the Mascoutens headed into the Wabash Valley for the winter hunt in fall 1702.[28] They promised to bring back hides to sell to the French. However, as the French did not yet have the capacity to tan the hides, the Indians would have to do that themselves.

The Illinois and the Mascoutens and others who came to visit Juchereau were excellent partners for his project. Traveling through the Illinois Country in the 1720s, Charlevoix noted that the Illinois produced bison leather that was better than anything he had ever seen. "As to the hide," he wrote, "there is none better in the known world." As Charlevoix noted, and as Juchereau probably also assessed, bison leather was perfect for military applications, a fact illustrated by the Indians' own use of it: "The Indians make bucklers of it, which are very light, and which a musket-ball will hardly pierce." Most important, Charlevoix noted, was that the hide did not take long to prepare. As he assessed, "it is easily dressed."[29]

In fact, there was nothing "easy" about the Indigenous tanning process; that statement was wildly false. That said, the essential mechanism of this process of tanning differed in important ways from the factory-based European trade. Like many other groups, the Illinois tanners used brains and other organs from the animal to create the greasy substance at the heart of the process. Whether starting with a "green" hide freshly removed from the animal or one that had been previously salted, the process began with soaking the hide to soften and prepare it. Stretching it on a frame, the hide worker then used a flesher to remove all the fat, meat, and tissue from the skin of the animal.[30] This was immediately followed by dehairing, usually performed with iron or stone blades attached to a handle, known in the Miami-Illinois language as a *massapic8i* or *cacahagane* or *assacagane*.[31] Here was the most meticulous part of the process; failing to remove any of the membranes in the skin would prevent the brain solution from penetrating into that area of the hide, resulting in a hard and tough spot on the hide. After wringing the hide to remove moisture, the tanner had the "rawhide" ready for the brain solution.

Removed from the animal and boiled together with some lard, the animal's brain was reduced to a thick oil that could be applied by hand, then rubbed in and massaged to make sure the emulsifiers in the solution truly penetrated. Often this took several applications, and sometimes hide workers folded the hide into a kind of envelope to intensify the penetration. After that, the hide was smoked over a smudge fire to assist further in breaking down the membranes in the skin, a process expressed in the word *nit8nsat8e*, "I smoke a dressed, worked hide."[32] After a few hours of smoking came softening by hand, as well as final wringing to remove moisture, both aided by another application of smoke. Now the hide worker stretched and worked the hide over a firm object (known in the Miami-Illinois language as a *Pic8a-hig8ntac8i*) to soften it up.[33] Again, it was a laborious part of the process, which was then followed by a final smoking to add color to the hide.[34]

The process was far from easy, contrary to Charlevoix's estimation. If anything, it was actually more laborious than the European method of

Hide flesher-beamer that belonged to Chief William M. (Bill) Skye, Peoria, 1868–1923. Courtesy National Museum of the American Indian, Smithsonian Institution (2/489). Photo by NMAI Photo Services.

vegetable tanning, which revolved around a long soak. But here was a key difference: because the hide did not have to soak in the tannic acid for a long time, the brain tanning process could be completed in a matter of weeks, not months. Moreover, while vegetable tanning was limited by how many vats were available for soaking, many hides could be brain-tanned side by side. All of this is important for understanding Juchereau's tannery. Not limited by special equipment, the only constraining factors for production lay primarily—and simply—in how many workers could be dedicated to doing the laborious tanning process at one time.

When it comes to that question of labor, Indigenous hide production was artisanal, and still is.[35] Many studies elaborate this artisanal and artistic aspect of hide tanning, and much of the ethnohistorical literature about Indian hide-tanning begins with the premise that it was a specialized and individual practice. Inextricable from the process of hide-tanning was and is the meaning of animals as other-than-human persons in many Indigenous American cultures, a meaning which invested the craft with sacredness.[36] Tanning hides touched on the important social relations between people and the other-than-human persons or relatives embodied in the animals. Moreover, the craft and practice of hide production also reflected important social relations among humans themselves, as it was ordered according to an important division. In particular, women often possessed the know-how and the expertise to produce hides, and this work was considered a special part of their domain and power.[37] In most every traditional context, women did hide-tanning alongside their agricultural production, another female domain over which women exercised great control. In many cases control over hide production was an important component of the complementary gender order that undergirded the communalism of Indigenous societies. Hide-tanning had simply not been practiced with a maximizing industrial logic like the one that Juchereau's tannery project envisioned.

This is the reason Juchereau's tannery was so transformative, however. As the project began, Juchereau's trade goods injected a new logic into the production of hides in the Indigenous communities near the confluence. Pursued not so much as a means of accumulating wealth for themselves, trade instead was valued by Illinois men and women as part of a culture of status-seeking in which an individual's generosity—the ability to give away and redistribute goods—was a central imperative. So consider how this status-seeking interacted with the problem of labor. Contra Charlevoix, bison hide production was extremely hard work. One nineteenth-century observer of plains hunters like the Blackfeet estimated that a single woman could produce just eight hides in a year if working by herself using techniques similar to the ones that the prairie people used.[38] If they worked in groups, however, they could produce many more. For instance, one chief estimated his eight wives could produce up to 150 hides in a year by working together, that is, more than doubling their output.[39] If Juchereau's technological innovations were a long way off, then, here was an efficiency gain that surely would have mattered at the "tannery." Here Illinois households were presented with new incentives to continue the pattern that they had already established during the 1690s, integrating additional women as second and third wives into their hierarchical polygamous families to assist with the new labor burdens.

It turns out this was not just an Illinois-centric phenomenon, and we can best understand these dynamics if we put them in a broader context with other Indigenous societies participating in the bison hide trade across time. As several scholars have noted, the maximizing logic of the hide trade frequently skewed the gendered division of labor for Indigenous societies, throwing out of balance an older and complementary division of labor. In these societies not only did women's status decline, but their labor was organized in a much more instrumental way, featuring both polygamy and slavery.[40] Importantly, as anthropologist Judith Habicht-Mauche writes of protohistoric groups on the southern plains, a general logic seems to have altered gender relations in such circumstances: "Since women were the primary producers of wealth, in the form of processed bison hides, the control of women and women's labor became an essential feature of men's competitive, status-building activities." In this context, male heads-of-households had a strong incentive to bring more laborers in. "Sustaining [the] competitive system would have required a steady input of new sources of women's labor. Bringing women into domestic units from outside the local area, either as captives or as exogamous marriage partners, would have been one way to expand this labor force."[41]

This integration of nonlocal women into Illinois society was by now a well-established pattern, going back at least to the 1670s. The Juchereau tannery changed two things about these long-standing patterns. First, Juchereau's ambitious scheme and Louisiana's boosterism implied that the French demand for bison skins would be unlimited, giving Illinois households every incentive to maximize their quest for extra laborers. Perhaps more important, however, the migration that brought the Illinois and the Mascoutens to the area around the Wabash valley also brought them into closer proximity to foreign groups who had long been both enemies and the source of captives, people like the Chickasaws, the Pawnees, and the Padoukés. Importantly, these people were for their own reasons embroiled in a regional slave trade in this period, one oriented around distant markets in the Southeast, especially Carolina. Entering this space, the Illinois and the Mascoutens found themselves both willingly and unintentionally surrounded by potential captives and potential enemies. Trading with Juchereau, the Illinois and the Mascoutens may have gained access to gunpowder. All of these factors combined to produce a brief but consequential uptick in violence.

◆ ◆ ◆

To understand this moment in ecotone history we must begin with dynamics that stretched far beyond its boundaries, and we must start with the Chickasaws. Upon arriving in Louisiana, Iberville reported the activity of the Chickasaws, the most powerful group in the Southeast, between the Appalachian Mountains and the Mississippi River, who were engaged with the English in a "very large commerce," the nature of which was supposedly "distressing to all the Indian nations" of the Mississippi Valley: the slave trade.[42] Like the Illinois, during the early colonial period the Chickasaws had made a business of raiding nonallied villages, taking dozens of people as prisoners to sell at market. What made this different from the Illinois' business was probably the scale, since the market that the Chickasaws served in the early 1700s was massive: Carolina. As Iberville soon learned, slaves captured by the Chickasaws were marched up the Ohio and Tennessee Rivers, destined for Charleston, where they would embark for the Caribbean to join plantation gangs on British sugar islands, a market with a significant demand. In 1702 Iberville learned that English traders lived among the Chickasaws on a somewhat permanent basis, arming them with rifles and encouraging them to make slave raids in the area. In the first years of the Louisiana project, French officials contemplated how to "alienate" the Chickasaws from the English traders, urging them to realize that these trade partners only wanted to send "entire families to Barbados and Jamaica."[43]

The Chickasaws' enterprise proved a potential problem for the Juchereau tannery. Upon his arrival in the region, the priest St. Cosme witnessed the aftermath of a Chickasaw raid on the village of the Cahokias, near the location of modern St. Louis in the confluence. The result was grim. As St. Cosme surveyed the scene, the Cahokia villagers "were still devastated by the attack made on them by the Chickasaw and Shawnees." Visiting with the newly arrived Kaskaskia villagers, St. Cosme learned the purpose of the raid: to capture slaves. As St. Cosme put it, "They had killed ten men, taken nearly 100 slaves, as well women as children." For both the Jesuit and the Seminarian missionaries in the area of the confluence, a major agenda item was to pacify these slave raids.[44]

Quelling these animosities would not be an easy task. The arrival of the Illinois in the region in great numbers was certainly viewed by the Chickasaws as an unwelcome invasion. Meanwhile, evidence suggests the 1699 Chickasaw raid on the Cahokias was not just an unprovoked attack, but rather a retaliation for a previous Illinois raid on a Chickasaw village. In the winter of 1701–2, as he attempted peacemaking with the Chickasaws, Iberville learned that the Illinois villagers already possessed Chickasaw captives, so he sent an envoy of three Canadian coureurs de bois, along with two Chickasaw delegates, to the Illinois, "to demand [from them] the Chickasaws who are prisoners there and to warn them not to make any more war."[45] In fact, Iberville had a plan to isolate the Shawnees, another group of English allies, who he considered "the only nation to fear." Punishing the Shawnees for their alliance to the English, Iberville hoped to convince the Chickasaws and the Illinois, among others, to cease their slave raids against each other and instead turn to a new economic activity, including bison hunting.[46] Indeed, Iberville made this appeal far and wide in the Mississippi Valley, entreating potential allies to abandon the slave trade and instead turn to supplying the French with skins.[47] The English, Iberville said, "don't love anything but blood and slaves." By contrast, Iberville promised that the new Louisiana colony would provide "all sorts of merchandise in exchange for bison skins, deerskins, and bearskins—*these* are the only slaves that I ask for."[48]

When it came to the Illinois in particular, however, Iberville did not understand the relationship between the production of bison hides (the slaves that he asked for) and the slave *raids* that he lamented. Here was a fundamental reason why encouraging Indians in the confluence to cease raids against each other was a futile effort, especially at this time. Sources are quite scarce, but it seems clear that the Illinois continued their slaving expeditions in the early 1700s after arriving near the Wabash. Evidence suggests they captured or traded slaves from groups in the Lower Mississippi Valley, which may be why St. Cosme observed among the Kaskaskias "women and girls dressed like the Tonicas."[49] If these women were in fact captives, they may have entered Illinois via trade with groups like the Quapaws, based to the south at the junction of the Arkansas and Mississippi Valleys. Moreover, the Illinois were now even closer to the Missouri River, the well-established main thoroughfare for transporting slaves from the west, like the Pawnees, who were the target of almost annual slave-raiding expeditions by the Illinois since at least the 1670s. Writing a somewhat derivative report about the western Indians in 1710, co-intendant of Canada, Antoine Raudot, reported on the Illinois' massive slave-making activities in a chapter entitled "The Manner in which the Illinois Make War against the Pawnees." In addition, they also made smaller-scale raids in the winter, perhaps against enemy villages that were not so distant.[50]

If the Illinois and the Mascoutens almost certainly continued raiding and trading for slaves among the groups of the Mississippi, Arkansas, and Missouri Rivers, some of this activity probably proved especially fateful in the early years of the tannery. Likely unbeknownst to the Illinois, villages throughout the Southeast were in the midst of a scourge at the precise moment that the tannery began, what historian Paul Kelton calls the "Great Southeastern Smallpox Epidemic."[51] At first the most devastating impacts of this epidemic were surely felt by groups like the Creeks, whose condensed villages and frequent interaction with foreigners from Carolina created the means by which smallpox could easily enter and then spread. Importantly, it was the slave trade in particular that likely created both the disease ecology and the specific vectors for the disease to take devastating effect. Open warfare eliminated traditional buffer zones between villages when refugees from the violence gathered together, creating ripe conditions for transmission. Meanwhile, war parties introduced foreign captives into crowded villages, and these people frequently shared infections with their hosts. As the slave trade rippled through the Southeast, so too did the epidemic. By 1699 the epidemic reached its westernmost extent at some villages along the Mississippi River.[52] One of the most devastated groups was a village of the Quapaws in the Arkansas Valley.

The Quapaws were indeed overwhelmed, particularly at their town called Kappa located at the junction of the Mississippi and the Arkansas Rivers. Arriving in this village in the winter of 1699, the missionary Thaumur de La Source gave a grim report: "This fine nation . . . is almost entirely destroyed by war and sickness. It is a great pity." The Quapaws were so sick that they could not hunt, despite the fact that La Source arrived during the prime winter hunting season. La Source noted other groups impacted by this epidemic, but none as badly as those in the Quapaw village. As La Source's partner St. Cosme put it, "There was nothing to be seen in the village but graves."[53] This was a major event in the history of the Quapaws, and one reason why their numbers decreased from around five thousand to two thousand over the course of a single generation. But if the epidemic spread among them as a result of slave-raiding and slave-trading, as Kelton says, the disease ecology and disease vectors that caused the epidemic to reach the Quapaw village in 1699 ensured that it did not stop there. Two seasons later, the epidemic consequentially spread to the Wabash Valley.[54]

The story of this disease outbreak is central to understanding the role of the tannery in ecotone history. Almost immediately after beginning their work for the tannery, the Kaskaskias and the Mascoutens both suffered from smallpox. The Mascoutens were hardest hit. Settled on the Wabash River, they had resisted the Jesuits' initial attempts to missionize in their village, and were "not very much inclined to listen to the instructions of the Missionary." But then came a dramatic turn: "At that very time a contagious disease desolated their Village, and carried off every day many [Indians]." At this point the Jesuits saw an opening, and Marest reported in retrospect that the Mascoutens now became more receptive to Christianity, if only temporarily.

The turn in the Mascoutens' attitude should be read as a reflection of the severity of the disease. Quoting one of the suffering Indians, Marest reported on the impact of the smallpox at this moment, which caused the Indians to reach out to the Jesuit: "We are dead," said one Mascouten man to the priest. "Do not kill us all." In fact, according to Marest, smallpox killed half of the village's population.[55]

The same epidemic spread to the Kaskaskia village, although details about how the disease affected the Kaskaskias are sparser. What we do know is refracted through another Jesuit report on the progress of missionization. Recalling an event during his work among the Kaskaskias in 1704 or so, Marest remembered a notable conversion. It started when one Kaskaskia man "was attacked by smallpox, with all his family." Soon his wife and some of his children were dead. The man himself was "brought to the point of death" and was "scarcely able to walk" owing to the "painful malady." One child in the house went "stone blind" as a result of the disease, and other members of the family suffered similarly. For the Jesuits this was yet another chance to narrate the remarkable work of God in "softening their hearts" and opening the Indigenous Illinois to Christianity. For us it is evidence that the smallpox outbreak of 1704 circulated in the village of the Kaskaskias as it had just done among the Mascoutens. Marest referred vaguely to other "sick people who needed prompt assistance" while on the summer hunt in that same year. It may well be that Father Julien Binneteau succumbed from the same smallpox outbreak, which Marest attributed to exhaustion brought on by traveling with the Illinois during the summer bison hunt.[56]

In short, it is highly likely that the smallpox outbreak in the Kaskaskia village was as consequential as the one among the Mascoutens. If it was, there is an important reason why. Indeed, it is significant to note that this was probably the Illinois' first experience with smallpox; they had until now avoided this most deadly part of the colonial encounter. The reason for their good fortune was not genetic or indeed biological in any way.[57] It was historical.[58] While many Great Lakes groups had already experienced great outbreaks of disease, it was only in the 1690s that the Illinois suffered their first unnamed disease epidemic at the Grand Village, as noted previously (see chapter 5). If this reality was a consequence of the way the Illinois were living in the tight confines of the Grand Village, the 1704 epidemic of smallpox was similarly entangled with the specific circumstances of their lifeway at this time, shaped importantly by hide-tanning and slave-raiding. As recent historiography has emphasized, far from being preordained and inexorable forces of biology, diseases produced their worst effects when they worked intersectionally with other causes of mortality.[59]

In the Wabash River Valley the truly consequential driver of disease may well have been the slave system, the frequent and large-scale raiding that defines the period. If the Illinois had once been "offline" and remote from major disease outbreaks in high-traffic places like Green Bay, the new slave system now made them very much "networked," with their villages full of nonlocal women, foreign captives, Europeans, and Indigenous delegates coming annually to negotiate peace.[60] Meanwhile, the newly intensive economy of hide production, an economy requiring teams of laborers working closely side by side, may well have supplied a ripe disease ecology and helped to make the disease extraordinarily efficacious. This supposition is, of course, partially speculative. But remember: the Quapaws were suppliers of slaves to the Illinois. Disease was, on one hand, a product of colonialism, but in the context of the tannery it may also have been a *by-product* of the economy featuring bison hunting, bison maximization, and captive raids that supported it. In other words, disease shaped and was also shaped by older trajectories in the lives and histories of Indigenous peoples that continued through colonial contact.

If all this is true, it is not hard to see why these events quickly became so transformative for the Illinois and the Mascoutens and generated the feedback loops and entangled cycles of cause and effect that linked violence and ecology in this moment. In the early 1700s, just like a generation earlier, losses from disease and battle only inspired ever more conflict and created the urgent imperative to restore balance to households and kinship lineages. In the wake of epidemics the Illinois and the Mascoutens found enhanced motivation to embark on slave raids and double down on the violence, since both could restore their households and increase their productive capacity in the trade. Armed with gunpowder from Juchereau's project, they may have mustered a more formidable military force against their western enemies. Exogenous and Indigenous forces entangled: here, indeed, was a uniquely intense engine of destruction.

And then, in an instant, the situation grew even worse. Sometime in the winter of 1703–4, Juchereau himself became sick and died. News of his death arrived in Louisiana in fall 1704. By now Iberville's brother, Jean-Baptiste Le Moyne de Bienville, was firmly in control of the fledgling colony, and he viewed the Wabash project with new eyes. As he saw it, the project suffered from a fatal flaw: the disorder of the coureurs de bois. As Bienville wrote in

1704, as many as 110 unlicensed traders were roaming in the region around the Wabash and Missouri Rivers, including many of Juchereau's original party. Bienville recalled them all, ordering them to build large boats and gather as many hides as could be transported to Louisiana. Establishing order was more important than increasing profit. The tannery was suspended.[61]

The end must have been as dramatic as the beginning two years before. As Pénicaut recalled in his account, the men at the Wabash "loaded [the boats] with more than twelve thousand buffalo hides," which must have been an amazing sight and an even more extraordinary load to manage. Approaching 120 tons of leather, here was a physical embodiment of an amazing assemblage of *work*: solar, photosynthetic, animal, and human (and especially female).[62] Evidence suggests that the hides never even made it downriver, and no French profit was ever realized from the intense but short-lived project.[63] Over the ensuring years Juchereau's brother, Louis de Saint-Denys, would take charge of the tannery operation, although with a new regional understanding. All that remained of the tannery in the Wabash Valley was a memory and the mention of an "*ancient fort ruiné*" that appeared on French maps of the Mississippi Valley over the next decades.

◆ ◆ ◆

Nothing went back to normal. Of course it could not. The bison boom had brought the Illinois and the Mascoutens and the Chickasaws and various neighbors together, setting them into competition within a volatile space and touching off new animosities. Meanwhile, the newly privileged access that the Illinois had gained to French trade rekindled old rivalries. Most importantly, the Sioux, who had been resentful of trading parties to the Illinois Country since before Juchereau's project began, jealously eyed the new project as a threat to their privilege at Fort Huillier. They made attacks and the Illinois counterattacked. The conflicts between the Illinois and Sioux kept the Illinois Valley volatile in the first decade of the 1700s, as Marest reported in his travel accounts quoted at the start of this chapter. The bison boom left the Illinois Country embroiled in violence in the first decade of the eighteenth century.

If it created immediate episodes of violence, however, the bison boom set up a longer-term atmosphere of crisis and uncertainty for the Illinois. Having moved in response to the Louisiana project, the Illinois may well have felt that they had made a losing bet. Louisiana was clearly interested in further development of the bison trade, but not anytime soon. For now, deprived of Juchereau's tannery, different Illinois speakers tried various and often competing strategies. Some Illinois like the Kaskaskias continued their lives in Christian missions, becoming ever more fervent Catholics and solidifying their alliance to the French. Some of the same adopted French farming practices and even took up animal husbandry, raising chickens and pigs.[64] Others took a different approach, rejecting the priests and French influence. The Peorias certainly continued to resist the French from their village in the mid-Illinois Valley, where in 1706 one Peoria grew so resentful of Father Gravier that he nearly murdered him. The rejection of the French also was strong among the decimated Mascoutens, where in 1702 Father Mermet had famously failed to convince an audience that the Christian God was superior to their own personal divinities, particularly the *bison.* Among the many Illinois speakers, the disorder of disease and the elusive profits of the bison trade did not fundamentally shake older worldviews and identities. Indeed, and in a larger way it is clear that Indigenous logics, and not any kind of dependency, continued to guide the actions of Illinois speakers in this moment.[65]

The best indication that the Illinois continued to follow their own priorities and avoid dependency shows when they turned their attention back to the north. Much had changed during the period of the Juchereau tannery, as Canada's government solidified its Great Peace at Montreal in 1701. Temporarily called away to a separate project, the Illinois had paid little attention to what the Great Peace entailed, especially the underlying vision of a pan-Algonquian alliance inclusive of their rivals the Meskwakis. Now that the tannery scheme had faltered, however, the Illinois had reason to focus new attention on the specifics of this Great Peace. In particular, in feeling the brunt of Sioux animosity, the Illinois had renewed motivation to set the Sioux against the Meskwakis to ensure that the two main enemies in the ecotone did not reconcile and turn against themselves. Rejecting the vision of 1701, in other words, the Illinois were about to target these groups and their newfound peace.

It is impossible to understand what came next without understanding the history of the bison tannery, however. As they inspected the situation in the north, the Illinois did so in the context of their recent experiences. In the wake of Juchereau's tannery, they had losses to recoup and kin to replace. Far from reducing them to dependency, the bison market had intensified the Illinois' long-standing ambitions for dominance in the midcontinent, even as it produced epidemics and violence that challenged those ambitions. The events in the Wabash Valley in the early 1700s had mixed the exogenous forces of colonialism—markets, diseases, and guns—with long-standing Indigenous dynamics like mourning war and captivity. Entangled with the bison, the Illinois had made a brief and stunning bid for power inside of this dangerous mix. Reeling from the consequences, they would now turn renewed ambitions and resentments back to their old ecotone rivals.

7

WAR

IN THE COLLECTIONS OF HARVARD'S PEABODY MUSEUM IS A 1735 watercolor painting by French surveyor Alexandre de Batz, surely one of the most valuable French eyewitness depictions of Illinois Indians from the eighteenth century. Painted near New Orleans a few years after the climactic battle of what became known as the Fox Wars, the image tells a story about that recent violence. When de Batz made the painting, an army of Algonquian and French soldiers, led primarily by the Illinois, had just finished routing their enemies the Meskwakis in one of the most extraordinary military encounters of early America, in the prairies near present-day Arrowsmith, Illinois. Now here they were on a journey to Lower Louisiana, celebrating and asserting their power. In de Batz's painting three young Illinois men wear mataché, the tattoos and other body art displaying what were likely trophies earned for specific acts of valor on the battlefield. The victors show off their weapons and dance in celebration, perhaps sharing a calumet ceremony with local Indians from Lower Louisiana as their chief looks on approvingly.

In addition to the striking depiction of the Illinois fighters, a key part of de Batz's 1735 scene—and the detail that makes the image such an important commentary on the Fox Wars—is a figure seated in the foreground. There in the corner, positioned submissively below a man identified as a chief, is a Meskwaki woman, classified as a slave (*Renarde Sauvagesse Esclave*). Historians have not ignored her, and they have not ignored the role of slavery in

Desseins de Sauvages de Plusieurs Nations. Illinois Indians at New Orleans by Alexandre de Batz, 1735. Watercolor. Gift of the Estate of Belle J. Bushnell, 1941. Courtesy Peabody Museum of Archaeology and Ethnology, Harvard University, 41-72-10/20

bringing on the Fox Wars.[1] Yet while recognizing the presence and *symbolic* meaning of captives like this Meskwaki woman, and the resentments and conflicts they inspired and embodied, historians have frequently ignored what captives like her were actually *doing.*

Look closely at de Batz's image. The Meskwaki woman sits amid a cargo of trade goods, helpfully labeled by de Batz. One box contains *plat côté,* or rib meat. Two adjacent barrels contain *suîfs,* or tallow, that is, rendered animal fat. Most importantly, look in her hands: she is working (or at least holding) a tanned hide. Aside from a package of "bear's oil," all of these commodities came from one kind of animal: the bison. They were products brought by the Illinois for trade in Louisiana, and they were almost certainly the products of the Meskwaki woman's labor. Here is the important and ignored meaning of de Batz's painting of the Illinois, and, by extension, his commentary about the Fox Wars. The picture, like the Meskwaki woman's captivity, like the story of the Fox Wars itself, is partially about the bison economy. Appropriately and prominently positioned right in the foreground of the picture, the enslaved Meskwaki woman suggests the connection between bison work, captivity, and warfare in this violent period of the ecotone's history.

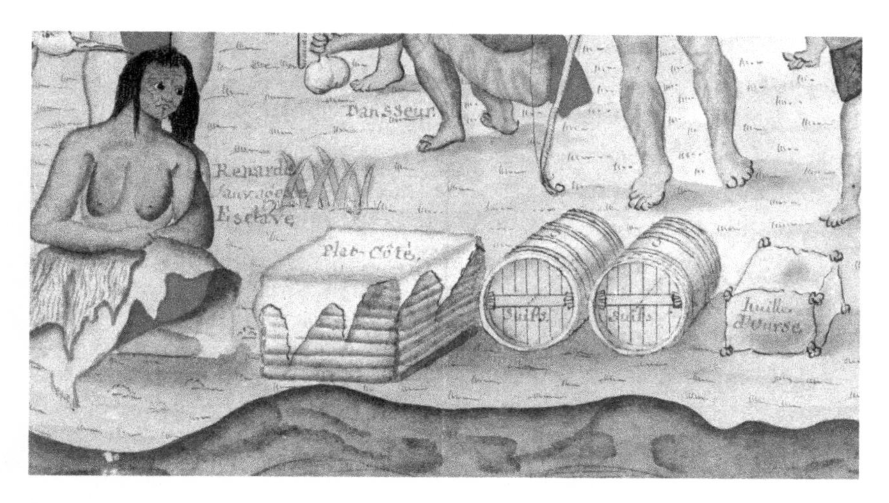

"Renarde Sauvagesse Esclave (Fox/Meskwaki Indian slave woman)" (detail). *Desseins de Sauvages de Plusieurs Nations* by Alexandre de Batz, 1735. Gift of the Estate of Belle J. Bushnell, 1941. Courtesy Peabody Museum of Archaeology and Ethnology, Harvard University, 41-72-10/20.

Following de Batz's lead, we have an opportunity to understand this complex story, shedding new light on these conflicts, which have often seemed so difficult to explain. This new light will make our narrative of the Fox Wars much richer. As they have attempted to explain the momentous events of the Fox Wars, historians have told incomplete and starkly divergent stories. Following French sources and their inflated sense of colonial control, for instance, some early chroniclers took for granted that the Fox Wars were the result of French planning and strategy, centering *French* and *colonial* actions as the primary drivers of the Fox Wars.[2] As a counter to these traditions, meanwhile, other historians have centered Indian motivations as the more important factors in the prosecution of the war. Locating the salient and driving agendas behind the conflicts in the Indian rivalries and the aftermath of the Beaver Wars, such historians have shown how the conflicts spiraled, especially as Indians made the colonists embrace *their* animosities and do *their* bidding on the battlefield.[3] Taken together, these different traditions leave us with a picture of the Fox Wars as a manifestation of Indian dependency *and* Indigenous power, an episode of genocide but *also* of colonial weakness.

The reality is that all of these previous stories contain elements of truth, and the seeming tension between their different emphases need not be resolved. Even as we recognize and center the agency of both colonists and

Indians in the Fox Wars, however, the story of the conflicts is incomplete unless we appropriately center the other agents that helped shape and drive them: *nonhuman* agents. Deeply entangled in Indian actions during this period were important material and economic realities, particularly the competition for trade and thus power by the interior groups who were the central participants in the conflicts. In the case of the Illinois, the central antagonists of the Fox Wars from beginning to end, the most important material circumstance was indeed the bison; the bison was the center of an economy that had transformed life for generations and that had recently become the hoped-for path to power at Juchereau's Wabash tannery in 1702. Although not usually featured in the story of the Fox Wars, the bison—and the power the animals represented—were central agents in bringing about the wars. More than just the animal, furthermore, it was the bison's specific history in the ecotone and the ecotone's history that mattered.

Recalling the lessons of previous chapters, we are now in a position to understand why. When the Illinois doubled down at the tannery in the Wabash Valley, it did not go as expected. When the project suddenly stalled in 1704, the Illinois turned their attention to the north and their historic enemies, the Meskwakis. Harboring animosity against this group that was rooted in a long history of competition, the Illinois now released it in familiar fashion: through captive raids. Only now *all* the logics of that practice—from ancient notions of friendship and enmity to imperatives for kinship replacement, to more instrumental ideas about labor—came together with shifting alliance politics fueling extreme kinds of raids against the Meskwakis. Encouraged and exacerbated by French policy, by the Meskwakis' own strategies as well as those of other northern neighbors like the Ottawas and the Sioux, the Illinois launched and led the largest captive raid in ecotone history in 1712. They then formed a new strategy of power-building among a Missouri River world, a strategy that not only inspired renewed enmity against the Foxes but also represented a logical conclusion of their long, steady commitment to the bison resource dating back to the 1600s.

It is not clear how much de Batz really understood about the relationship between the Illinois' recent conflicts and the bison commodities that they brought to Louisiana in 1735. But his painting is a useful reminder that the economy was a big part of their actions in these years. For the Illinois, the bison lifeway, slavery, and warfare were now all of a piece. Inseparably, they drove one another, as the Illinois tried to rebuild and extend their power in the tallgrass.

◆ ◆ ◆

To understand the coming of the Fox Wars we have to start at the very moment Juchereau's Wabash tannery was falling apart. In 1704, when Juchereau died, another new French colonial project was taking shape at Detroit. Founded in 1701 by the legendary Antoine Le Mothe dit Cadillac, Detroit is often viewed as a quixotic scheme by a colonial dreamer, and it surely was that. At the same time, it was also a pragmatic and calculated response to a series of problems faced by New France at the turn of the century. In 1697 New France had suspended the western fur trade in response to a massive glut in beaver production that threatened to drive prices down below profitability (for the colony, if not for individual traders). Four years later, however, the governor of New France, Louis-Hector de Callière, finally sealed a long-anticipated alliance among the Great Lakes Algonquians at Montreal, the so-called Great Peace of 1701. These developments were in no small degree of tension. To maintain the alliance of 1701, New France would have to overcome constant competition from English rivals. And it was through the fur trade, on terms advantageous to Indian partners, that the French would need to secure and preserve friendship with their Algonquian allies.[4] Detroit would advance these causes.

In its conception Detroit might be viewed as a kind of mirror image of the grand plan that Louisiana governor Iberville had enacted for the outpost at the Wabash. Like Iberville's, Cadillac's plan featured similarly elaborate population engineering, including encouraging entire populations of Indians to relocate to a French-chosen location. Newly transplanted with their permanent villages just a short distance from the French colony, these groups would become the major suppliers of furs at the new post. They would also accept missionaries to instruct them in Christianity. Most important, they would make peace together. Soon Detroit would rival or even eclipse previously established trading centers—Michilimackinac, Chequamegon/St. Esprit, Green Bay, and Fort Saint Louis in the Illinois Country. It would be a bulwark of French trade and diplomacy, a solid foundation for the perpetuation of the 1701 alliance.[5]

To achieve his plan, Cadillac hoped to attract the most important beaver-producers of the region—the Ottawas, the Hurons, and other groups besides—making Detroit a primary trading center for people of the western and southern pays d'en haut. The new Detroit marketplace would effectively circumvent the northern trade routes west of Michilimackinac, where hostilities among the Sioux, the Meskwakis, the Ho-Chunks, and other groups frequently disrupted and made the trade dangerous for French traders. But here is why the Detroit colony was so controversial. Insensitive to the ways his new scheme disrupted the western tribes' status quo, Cadillac imagined Detroit as a way to achieve a simple vision of pan-Algonquian unity. When it came to the Meskwakis in particular, Cadillac included them without qualification, ignoring the resentments that the Meskwakis' participation at Detroit might trigger among the Ottawas and the Sioux. He also welcomed the Mascoutens and the Kickapoos, allies of the Meskwakis, making privileged room for them just on the outskirts of Detroit as well.[6]

For no group was this vision of pan-Algonquian unity less agreeable, however, than for the Illinois. The ecological underpinnings of the Illinois' renewed resentment were implicit but important. Trade at Detroit promised renewed opportunity—and power—to beaver hunters, creating a new market in this valuable commodity of the northern ecotone. And where the Illinois' inferior beaver pelts had caused them to focus on the future of the bison trade with the French (whether at the Wabash or Detroit or anywhere else), their rivals, including the Sioux and the Meskwakis, lived in a more northern environment that allowed them to supply both animals. In this context it was to the Illinois' advantage to keep the Sioux, the Meskwakis, and others divided and antagonistic to prevent them from amassing power. Of course, that power was no sure thing; for instance, it was not easy to unite the lakes tribes at Detroit, as the French found out. In 1712 as the Detroit colony was taking shape, the Ottawas and the Potawatomis attacked each other and then united against the Kickapoos. Several Algonquian residents at the new outpost eyed the Meskwakis warily. In 1712, as the Detroit colony was taking shape, the new residents of Detroit jockeyed among themselves for advantage.[7]

If several northern Algonquian groups and the Hurons questioned the Meskwakis' inclusion at Detroit, it was the Illinois who most firmly rejected it. This is why in the spring of 1712 an Illinois chief named Makouandeby led a large army of allies up the Illinois Valley to Detroit to protest against the Foxes' inclusion. As the now-commandant at Detroit, Jacques-Charles Renaud Dubuisson, recalled, Makouandeby's massive and unexpected expedition, comprised of "the army of the nations of the south," arrived to attack the Foxes. Dubuisson later wrote, "They were the Illinois, the Missouris, the Osages and other nations yet more remote"—in other words, all bison people. Meanwhile, these tribes organized a coalition with lake tribes who shared their resentment of the Foxes. As Dubuisson put it, "There were also with them, the Ottawa Chief, Saguina, and also the Potawatomies, the Sacs, and some Menomenies." It was unprecedented: "Detroit never saw such a collection of people. It is surprising how much all these nations are irritated against the Mascoutins and the Ottagamies [Foxes]."[8]

While surprising to Dubuisson, perhaps the anti-Fox "irritation" felt by members of this army of "nations of the south" and from regions "yet more remote" should not be so surprising to us. Marching "in good order," this unprecedented army arrived to both insist against the peace-making and policy changes that Detroit was founded to enact among the northern groups and to isolate the Meskwakis. Calling them "dogs" and framing the Illinois as "masters," the Illinois chief Makouandeby spoke directly to the Meskwakis in a metaphor of slavery, revealing his intention to subordinate them even as he rejected the Foxes' inclusion at Detroit: "Go away, then. For us, we will not stir a step from you . . . We shall see from this moment, who will be master, you or us; you have now only to retire, and as soon as you shall re-enter your fort, we shall fire upon you."[9]

Makouandeby's speech powerfully expressed the Illinois' desire to reduce their rivals. Acknowledging explicitly that the Illinois' intervention at Detroit "disobey[ed]" the intentions of "the French Father" (the governor of New France), the Illinois knew they were provoking a special turning point in their relationship with the Meskwakis and with the colony. Makouandeby's question of "who will be master" had of course not been far from the Illinois' minds over the previous two generations. Now the Illinois chose "this moment" to release their pent-up resentment against the Foxes, the "wretches" whose move to take advantage of the Detroit opportunity came just as the Illinois' own path to power—the tannery—seemed uncertain.

It is impossible to exaggerate the importance of Makouandeby's determination—and the Illinois' role—in what happened next. We cannot take at face value retrospective reports from the likes of Dubuisson or Governor Vaudreuil, who later positioned the French as "the first movers in this war."[10] Instead, we need to remember whose agenda the war truly served and that it started with the Illinois "disobey[ing]" French intentions, as Makouandeby's speech put it. In besieging the Fox village and chasing its residents away from Detroit, the Illinois and their partners launched the most significant slave raid in ecotone history. Their effort took massive casualties. In the dramatic final stand near Gross Pointe, the anti-Fox army led by the Illinois and the Hurons slaughtered hundreds. The Foxes lost a thousand souls, about one hundred men killed and some nine hundred women and children taken prisoner.[11] Though we cannot know the precise number, we know from many later reports that one group in particular took an especially large share of the prisoners from the battlefield. When the Illinois returned home in the fall of 1712 they brought with them a large caravan of Fox captives, numbering not in the dozens but probably in the hundreds.[12] This was mourning war on a massive scale.

◆ ◆ ◆

To understand the rest of the Fox Wars we need to keep our eyes on the slaves of 1712. They appear in French writings and Indian speeches many times, and they drove a generation of conflict that followed the events at Detroit. The Fox Wars, of course, involved other actors, and its vicissitudes are endlessly complex. Yet the resentments between the Illinois and the Foxes regarding these captives was frequently a central impetus for the fighting. As naive as they often were about the Indians' true motives, even the French could perceive at various moments that peace might reign among the Algonquians if only these Illinois-speakers would return their Meskwaki slaves from 1712 and allow others to do the same.[13] These captives drove the course of conflict.

It was an unpredictable course of conflict to be sure. Eager to have peace in order to restart the fur trade, French administrators in Canada at first hoped to mediate the fighting and stamp out any lingering resentments and restore the hoped-for "unity" of 1701. Initially they tried to treat what happened at Detroit like a normal episode of violence, pressuring the antagonists to "[make] presents for Covering the dead, and of Recovering the Slaves that may have been taken on either side." But they quickly realized that neither the massive 1712 battle nor the Foxes' retaliation in its aftermath were normal.[14] Here was the origin of the language of extermination in the story of the Fox Wars. Changing their policy, the French planned an elaborate military expedition by organizing several Indigenous armies from the Great Lakes for a massive assault to subdue the Foxes and end the fighting, if not also the Meskwakis' very existence. This expedition, scheduled for the summer of 1715, would be a show of force so overwhelming that it would either "exterminate" the Foxes or reduce them to a stable peace. As one French planner put it, the goal would be "peace or war, as may seem to the purpose."[15]

The Illinois must have viewed this French policy as aligned to their own purposes. They were enjoying a moment of strength in the wake of their victory at Detroit in 1712, looking forward to new prospects. After the closing of the Wabash tannery in 1704, Louisiana officials now discussed "reestablishing" the project where "the Illinois [Indians] would furnish a great quantity of [bison hides]." More and more French traders, some of them remaining from Juchereau's outpost, settled in the Illinois Country among the villages of Kaskaskia and Cahokia, an optimistic sign for future trade. Should Louisiana's economy fail to develop, meanwhile, English traders arriving in the Wabash Valley from the east in the 1710s represented new possibilities. In particular these "English of Carolina" began building "three forts" on the Wabash, increasing the Illinois' access to that colony's robust slave market, to which the Illinois may have traded at least some Fox slaves after 1712. It was partially in response to this threat that the French in Canada, determined to align their own policies with what they realized was the Illinois' priority, to "carry out *their plan* of making war on the reynards" (emphasis added). The Illinois periodically sent Fox slaves to New France as gifts to symbolize the Illinois' power and to wedge the Foxes further from Onontio. Here was a reality—like the slaves themselves—that Canadian officials had to accept.[16]

The French were happy to accept these slaves, just as they were happy to contemplate violence against the Meskwakis in 1715. After all, the French wanted nothing but to have peaceful trade, and for that they were willing to allow what they saw as the Illinois' and other allies' priorities to drive their policy. But then in an instant another reversal clouded the plan and changed the French calculation. As New France officers traveled through to organize the expedition of 1715, epidemics struck Illinois and Miami villages in the Wabash and Illinois Valleys. Likely the measles, these were some of the most "pernicious" outbreaks in the history of the region. Among the Miamis "there were from fifteen to twenty deaths a day." Around two to three hundred Illinois succumbed at Kaskaskia. The long-serving Jesuit Gabriel Marest died in the epidemic. Illinois participation in the 1716 expedition now looked much more uncertain. Reeling from sickness, the Illinois joined together to attack a more modest target—a village of "70 Cabins of Mascoutins and Quikapous, allies of the Renards, who were hunting along a certain river" near the Illinois Country. They were successful, killing "more than 100" and "carry[ing] away 47 prisoners, without counting the women and children." Now the Illinois felt satisfied, with the new captives having "restored their spirits."[17] When the French finally set out from Chicago with their badly diminished army, the Illinois stayed home.

All of this was consequential. When Louis de la Porte de Louvigny traveled to the Fox Village at Butte des Morts on the Fox River, Louvigny's army was smaller than expected and the Meskwakis had dug in defensively. The Meskwakis' military strength even included some three thousand women armed and ready to fight. Unable to defeat them as dramatically as they had done at Detroit in 1712, the Indigenous army with Louvigny's help merely brought the Meskwakis to terms. If in 1713 the French had perceived that "all the nations of the lakes and inland tribes are purposing to destroy [the Foxes]," the outcome here was nothing so dramatic.[18] In this context the pragmatic Canadian French thus changed the purpose of the 1716 expedition and eventually made a new peace with the Foxes in 1716 at Montreal.

◆ ◆ ◆

To understand the impact of the peace of 1716 for the Illinois, two pieces of context are necessary. First, after years of glutted markets, factional disputes among French companies, and violence in the upper country, by 1716 the beaver market had seen an unprecedented upswing and become the best market in a generation.[19] Meanwhile, months after Louvigny signed the treaty at Montreal, the king decreed that the Illinois Country—including the Illinois Valley and points south—would officially become part of Louisiana. This was a major decision. Starting with La Salle's initial explorations, Louisiana's economy was perceived by New France officials as competition to their interests and a threat to their trade. With the 1717 decree, this old perception was validated by official policy, as French traders in Illinois—and in many ways the Illinois themselves—were placed in a different jurisdiction. Here, then, was a new line in the ecotone—a line of intra-imperial rivalry. Sanctioning through imperial policy what was already true in ecology, the French policy put the Illinois in a separate category. As it planned its promising new beaver trade, Canada was indifferent to—if not hostile to—the economic fate of the Illinois outposts.

In this context, the 1716 treaty strained the Indigenous Illinois' relationship to New France considerably. By resolving conflicts among the Foxes, the Sioux, the Ottawas, and the Chippewas, the 1716 treaty gave New France's traders new freedom to travel in what had long been dangerous territory. But the more important freedom it granted was surely to the Meskwakis themselves. No longer fearful of rivalries with their neighbors, the Foxes could potentially direct their rivalry against the Illinois. And, given the Illinois' position in Louisiana jurisdiction, New France officials had no particular incentive to stop them. The treaty threatened to isolate the Illinois.

Meanwhile, even worse for the Illinois than the peace itself were its terms. As Vaudreuil summarized the agreement, the Foxes promised "that they shall restore or cause to be restored all the prisoners, of every Nation, whom they hold." Were the Foxes to do this, the treaty implied, they could then reasonably demand that their former enemies would do the same. Even the French would have to contemplate returning their own Fox slaves. This was significant. As one wrote, "to make peace, it is necessary to begin by restoring to the Renards all the slaves of their nation whom the French hold; it is not in nature to think that peace can be made with people whose children we are withholding."[20]

The expectation of captive repatriation at the heart of this treaty disadvantaged the Illinois. It was asymmetric—after 1712 the Illinois held many more slaves than the Meskwakis did, and the latter in fact purportedly restored their own captives *before* the treaty was even signed. Worse, the prospect that others—and especially the French themselves—would repatriate the slaves that Illinois fighters had gifted them in order to wedge apart the alliance and create obligations was similarly abhorrent. Worst of all, however, was a final clause in the treaty. Recognizing the value of captives inside Algonquian villages and families, the treaty proposed a way to compensate for all of the redemptions that the treaty encouraged: "[The Foxes] shall go to war in distant regions to get slaves, to replace all the dead who had been slain during the course of the war." This was important. Ever since Marquette first visited the Peorias, and perhaps even long before that, the Illinois had used slaves as a commodity to overcome deficient beaver supplies, capitalizing on their ecotone geography to supply needed captives from the west to neighbors in the Great Lakes. The 1716 treaty now put the Foxes in direct competition with the Illinois for a slave trade that the latter had long dominated in "distant regions," namely the Missouri Valley. In this context the Illinois must have rejected the whole treaty as contrary to their interests.[21]

In the face of the peace of 1716, the Illinois held onto the Fox slaves from 1712. They were a symbol, legible to the Indigenous peoples of the ecotone and to New France and Louisiana officials alike, of the Illinois' rejection of the peace. But they also remained important in a material sense. After the epidemics of 1715, the Illinois had a growing need to replace relatives. As for their economy, their novel position inside Louisiana's government was about to include an official French commandant and garrison, both harbingers of renewed trade near their villages. Thus the captives remained valuable on many levels. For the Illinois, all of the incentives of the moment seemed to line up against accepting New France's mediation and in favor of more war in the ecotone.

◆ ◆ ◆

After 1716, the Meskwakis reveled in the friendship and influence that they had never previously enjoyed among the *nations des lacs*. They renewed their connection with the Iroquois, enjoying "some sort of alliance with them" that strengthened their power. More importantly, with the peace of 1716 and renewed access to trade goods, the Meskwakis now enjoyed control of the trade route into the western Great Lakes. With most of the Sioux fur trade passing through Fox villages, the Foxes leveraged a new alliance with their former enemies that became official in 1721. Apart from the Iroquois and the Sioux, the Foxes reached out and settled old scores with the rest of "the lake tribes" with whom they had previously fought—Ottawas, Chippewas, and others. Previously blocked from the profits of the beaver trade by long-standing enmities, the new peace gave them an unprecedented path to power, just as they had once hoped to achieve at Detroit. Remarking in 1719 on their significant reversal of fortune, Vaudreuil wrote: "The Renards have forgotten the sorry State to which they found Themselves reduced four years ago." Of course, the French eyed them warily: "The peace that was granted to them has served only to increase their pride and their Insolence."[22]

The Illinois eyed the Meskwakis even more warily than the French did. Unsurprisingly, given that they still held so many Fox captives, the Illinois were the target of much of the "insolence" that Vaudreuil had identified in 1719. In 1718, Fox and Kickapoo armies attacked Illinois villages. The Illinois counterattacked. While hunting bison on the prairie near the Rock River in northern Illinois, one Illinois party surprised a mixed camp of Foxes, Kickapoos, and Mascoutens, killing over twenty victims and taking captive several women and children. But the Foxes probably got the best of the Illinois in these years. We know from later accounts that the Foxes continued repeated attacks against the Illinois through 1719, taking many prisoners in at least eight separate raids, including one attack on Peoria in which Fox attackers burned fifteen and killed an additional thirty.[23]

Despite losses like these, the Illinois' decision not to surrender their captives and seek peace reflects their hope that they could capitalize on their unique situation. The biggest reason for optimism was Louisiana. The Illinois surely had watched expectantly as Louisiana sent repeated expeditions, beginning in 1714, to scout possible sites for a new trade center in the prairie-plains. For instance, Louisiana's new governor, Cadillac, had sent Juchereau's brother to the Red River in 1713, and he also sent a former coureur de bois, Étienne de Veniard de Bourgmont, on a journey up the Missouri River the following year.[24] Closer to home, Claude Charles du Tisné had arrived on an exploration of the Wabash in 1714, scouting for a possible replacement for Juchereau's outpost. While the first two of these expeditions eyed economic and strategic prospects beyond just the fur trade, including strategic alliances around Spanish outposts in New Mexico, they were also efforts to restart the ambitious projects put forth by Louisiana visionaries like La Salle and Iberville. In 1719 Pierre Duqué de Boisbriant founded the first official provincial government in Illinois Country, putting an outpost of Louisiana's administration near the center of the Illinois' villages at Kaskaskia. It was a new era representing renewed hope, especially for trade.

Most promising were Bourgmont's activities on the Missouri. Bourgmont himself had been present at Juchereau's tannery in 1704, and he brought back bison hides after exploring the Missouri in 1714.[25] When du Tisné reascended the Missouri in 1719, following in Bourgmont's path, he established contact with bison hunters like the Missourias, the Osages, and the Pawnees. To the Illinois, who exercised considerable influence with groups in the Missouri Valley, all of this activity—including most of all Boisbriant's new government near Kaskaskia—seemed to promise the reestablishment of trade in the Illinois' particular commodity.

The Illinois had to position themselves strategically among regional powers. Boisbriant's goal clearly was to make a regional peace in Missouri Country among the Pawnees, the Osages, the Missourias, and, especially, the Padoukés. This intended peace was part of du Tisné's mission in 1719, and it was the goal of another expedition launched by Bourgmont in 1724, when he ascended the Missouri River for a second time. Here was a possible strategic opening for the Illinois. To be sure, accepting French mediation in the west meant losing a source of captives, especially Padouké and Pawnee slaves.[26] On the other hand, peace among the bison people might encourage renewed export of their special commodity. In 1723 Bourgmont founded Fort D'Orleans, intending it as a permanent center for alliance-building and trade on the Missouri River. Significantly, when Bourgmont organized a delegation of Indians to travel to Paris the following year, the Illinois sent Chief

Checagou alongside chiefs of the Missouris, the Otoes, and the Osages, all fellow eastern prairie bison hunters, to advocate for this regional development. Their journey to Paris indicated a hopeful appeal for peace, for help against their Meskwaki enemies, and for future trade.[27] The Illinois were including themselves in this new regional bloc of Missouri Valley bison hunters.

As they contemplated this future peace and trade in the west, however, the Illinois now only had renewed reason (and power) to block and isolate their rivals, the Meskwakis. For one thing, if peace in the west were to end access to Padouké and Pawnee captives for the Illinois, the Meskwakis could remain a useful source of captives. Meanwhile, isolating the Foxes from their northern neighbors kept the Foxes on their heels, preventing them from developing their own power in the Missouri Valley world, where they had tried to establish connections.[28] For the Illinois, continuing to isolate the Foxes and building power in the west were of a piece.

With all these developments, however, now the Foxes continued to see the Illinois—and the Illinois French—as their strategic enemy. Certainly they attacked Illinois villages more, almost surely with the tacit support of Canadian officials. Regarding this fighting, in 1719, New France governor Pierre de Rigaud, Marquis de Vaudreuil complained that the Illinois were to blame since they continued to hold Meskwaki slaves from 1712 and to wage war "without any Regard for the action of the Renards," who reportedly had repatriated all their Illinois slaves. The Illinois only needed to reciprocate to calm the fighting. In Vaudreuil's telling, the Foxes "themselves were only desirous of living at peace with all the Nations," if only the "officer in command among the Ilinois [would] induce that Nation to make overtures to obtain it."[29]

The violence reached a crescendo around 1722. Traveling through the country around that time, Father Charlevoix saw signs of vengeance everywhere; for instance, dead Meskwaki bodies were hanging on scaffolds, where the Illinois had hung and tortured them. In response, the Foxes surrounded the Illinois at Starved Rock—notably near the location of the Grand Village, the center of Illinois power a generation previous—in a massive attack. Briefly the Illinois sued for peace, promising to send back all their Meskwaki captives. But they did not follow through; we know that the Illinois continued to take and hold new Fox slaves in subsequent months. In 1722 a French traveler observed a Cahokia war party returning to their village with recently captured Fox slaves. In this moment, and surely in response to these events, the Illinois gained a special reputation in the French imagination. As Charlevoix wrote, there were "perhaps no Indians in any part of Canada with fewer good qualities and more vices." Viewed by many as vengeful and belligerent, the Illinois were exercising power, but they were isolated and excluded from New France. Acknowledging both their power and their strategic isolation, Charlevoix noted that, "the Canada tribes . . . despised them heartily, but the Illinois were not a whit less haughty or self-complacent on that account."[30] In New France's eyes, the Illinois stood alone.

◆ ◆ ◆

Despite the violence, life in Illinois continued. Perhaps the best view into what the Illinois were doing at this moment was in the account of Diron D'Artaguiette, a trader and inspector for the new Louisiana government. Traveling to Illinois in 1722, he wrote with an understanding of the Illinois' situation and their ambitious activity inside Louisiana's economy. Traveling between Kaskaskia and the Ohio River, D'Artaguiette noted both sides of the Mississippi "lined with bulls and cows" gathered in herds of a hundred strong. As D'Artaguiette put it, "the most worthless Frenchman can kill a buffalo in this region." Of course, for the skilled Illinois hunters, it was possible to kill many more. And as D'Artaguiette's account makes clear, the Illinois now had begun to supply a new market for bison meat, particularly among the troops newly arrived to staff the new French outpost, Fort de Chartres. The cured ribs that the Illinois produced from the bison for this purpose became a critical ration for soldiers, and they were in high demand. Additionally, the Illinois also "traded in skins, such as beaver, buck, and deer, buffalo and bear."[31] Traders now came from both Louisiana and Canada (in defiance of regulation) and brought what was clearly a steady and considerable supply of merchandise.

In D'Artaguiette's estimation, all of this allowed the Illinois to preserve and even build power amid the violence. Although D'Artaguiette, like many Frenchmen, viewed the Illinois as still partially reeling from long-ago losses to the Iroquois, he nevertheless recognized their dynamism: "The Ilinnois are in general the handsomest and the best built [Indians] that I have seen." They had a distinctive material culture: "They clothe themselves and also their women with buffalo skins, which they dress on the flesh side and leave the hair which is long and fine." When trading with the French they also produced "porcupine work, which is very well known in France (where a good deal of it has been sent)." Some women carved out autonomy in trading,

D'Artaguiette observed, exercising control over their commodities. And yet, importantly, D'Artaguiette hinted at the role of slavery in all of this economic activity. The clearest sign was in the way that households and families were organized within the Illinois villages. Not only did women perform all the farming, as well as "dressing deer and buffalo skins"; the unequal division of labor was joined by what D'Artaguiette perceived as great tension. "The husband has full power and authority over his wives," he wrote, "whom he looks upon as his slaves, and with whom he does not eat." As we have seen, the reason Illinois men "looked upon" some of their wives "*as* . . . slaves" was almost certainly because a significant number of these women literally *were* slaves, including probably significant numbers of Meskwakis. Meanwhile, while Illinois and captive women worked hard to produce commodities for trade, status-seeking men continued to engage in their own role of production, which was different. "The men concern themselves only with hunting and making war," especially slave raids. They wore the mataché, which demonstrated their military accomplishments, "staining themselves . . . with red, yellow, black or blue." As D'Artaguiette observed, the entangled processes of economy and militarism continued to shape Illinois culture in the 1720s as they warred with the Foxes.[32]

Indigenous diplomacy after 1722 was at a low ebb—the calumet, for instance, was ineffective against the extreme violence. As Richard White puts it, continual wars among the Foxes and their enemies had left behind too many "ghosts"—missing kinsmen in need of replacement. Families naturally wanted to act to restore them. Against this urge, the whole idea of "tribal" unity and "chiefly" authority was no match. At the time a famous Meskwaki chief called Ouachala urged young Meskwaki men to follow the 1716 treaty and embark on attacks only against the Osages, rather than against the Illinois or other Algonquians. But there were many young men like White Buffalo, the hero of a Fox oral tradition from this period, who became famous for unstoppable vengeance against the Illinois. Indigenous diplomacy was no answer for White Buffalo's need; indeed, the logic of captive restoration was a continuing contributor to the violence. The same was true among the Illinois, where peace chiefs were not powerful enough to calm the animosity of young fighters. Describing them in 1721, a French observer noted that "the Illinois generally speaking do not recognize any chiefs. One is as great a master as another." In this context, the Foxes told the French that there was no chance of peace with the Illinois: "The Illinois have attacked us too often to allow of our staying our War-clubs."[33]

The problem became clearest when Canadian officials tried to mediate again in 1725, appointing a diplomat called Ouábessébau, or le Chat Blanc, for an extraordinary mission to the Illinois Country. Born to an Illinois mother, Ouábessébau was also brother to the chief of the Sauk at La Baye (Green Bay), and thus well-connected to the Meskwakis. In 1725 he went on a fact-finding mission to learn just what it would take to negotiate peace between the Illinois and the Meskwakis. At issue, most importantly, was the question of the Fox slaves from 1712: what had become of them, and could they be repatriated?[34]

Arriving in Illinois, Le Chat Blanc's fact-finding had to cut through the fog of war. On the essential question—did the Illinois hold Fox slaves or not—both the French in Illinois and the Illinois themselves issued denials. Replying to Chat Blanc, Illinois spokesmen Anakapita and Massauga responded rhetorically: "He says that his Slaves have not been given back to him. Where are they? Is there a single one in our villages? Does the renard speak the truth?" Turning the tables, the Illinois spokesmen insisted they were the ones who had been repeatedly attacked. Le Chat Blanc collected testimony from two Illinois, who provided a litany of Fox offenses, listing by name thirty-nine Illinois victims and referring to more than one hundred others lost in recent Fox attacks.[35]

French officials and priests in Illinois seemed to agree, detailing the attacks that the Foxes had perpetrated not just on Indian villages, but on the French settlements of Illinois as well. Supporting the Illinois' story of unprovoked Fox attacks, the priests pointed to recent episodes at Pimíteoui and Starved Rock. As for slaves, Commandant du Tisné insisted that "our Illinois have no Slaves belonging to the Renards, and have Never acted Treacherously toward them. They [merely] defended themselves." The answer to New France's question about the presence of Fox slaves in Illinois was clearly negative. Or *was* it? One Illinois priest was a little more ambiguous. Surely, he wrote, "in *our Illinois Villages* there are no renard slaves" (emphasis added). But since the Illinois were presently out on a bison hunt, "we do not Know whether there are any other slaves among Them."[36]

In fact, it seems obvious that there were Fox slaves among the Illinois in those other villages, as was widely reported and well known as far back as 1712. Vaudreuil had received reliable intelligence in 1724 that "the Illinois had not yet given any satisfaction to the Renards with regard to their prisoners, although the latter had sent theirs back to the Illinois." Constant le Marchand de Lignery, another New France official, echoed Vaudreuil, writing to Boisbriant in Illinois that "the Fox Are indignant because, when peace was made in 1716, they sent the Illinois back Their prisoners while The Illinois did not return Theirs, As had been Agreed upon in The treaty." Vaudreuil understood these "grievances" grew out of the fact that the Illinois "detain their prisoners." The denials by the Illinois did not hold up.[37]

From our standpoint centuries later, they do not hold up for us either. Too much circumstantial evidence suggests the prevalence of slavery in their society and the continued presence of Fox slaves in Illinois Country. Naive French eyewitnesses, for instance, still noted the way Illinois men treated their "wives" in their polygamous households, as in Lallement's account from 1721. "They are more jealous of their wives than the Spaniards," the traveler wrote. "On the slightest suspicion of infidelity they scalp them." As we have seen, such accounts detailed the treatment by Illinois men not of their native-born Illinois wives, but of out-group slaves. And while these wives may have been captured among western enemies, it is likely that many such slaves among the Illinois—in both the Indian and French villages—were Meskwakis. French priests baptized children born to Frenchmen and "red Fox slaves" in the French parishes at Kaskaskia and Cahokia.[38]

Given how Le Chat Blanc's fact-finding quickly descended into a dubious airing of grievances and denials, its value for history lies not in how it clears up specific questions about captives themselves but instead in how it reveals the Illinois' view of their larger situation and a clear and honest assessment of where their interests lay. As they now saw it, their interest lay primarily in the west, where they looked forward to increasing bison trade and power in the Missouri Valley. To achieve this dominance, however, they needed to foment division among their northern neighbors and continue the exclusion of the Foxes. In the Illinois' view, the carnage of recent years was caused by the large and unchecked Fox armies that were the logical outcome of peace among the ecotone's previously divided groups. Although New France officials valued that harmony, the Illinois did not; for them it was the *division* among groups that created an essential balance of power necessary for them to thrive. As one of the Illinois chiefs put it, it was better when the Lakes tribes were split: "The peace that was concluded at la bay [in 1716] has no other object than to harass us still more; for when [the Foxes] were at war with the nations They could not come in so Large numbers to destroy us." French officers in Illinois Country understood this logic too. Before 1716, du Tisné wrote, the Ottawas, the Sioux, and several others were "nations against whom the renards had to defend Themselves." With the peace, the Renards "have only the illinois to Contend with." The ecotone's essential quality—enmity—was thus gone, and the balance of power was ruined. The Illinois would have been better off without the mediation: "Had those Gentlemen let the tribes alone, we would have had to Fight against fewer Enemies."[39]

Even if the Illinois were clear about why division and not peace was the preferable condition, they also made an important point about the ecological underpinnings of their situation. As they put it, their alienation from New France's policy was centrally caused by a nonhuman actor: the beaver. The beaver had created an incentive for unity among the northern tribes, which gave the Foxes their newfound prosperity and wherewithal to attack, putting the Illinois outside of the alliance. As du Tisné put it, acknowledging the biophysical reality that made Illinois Country different from the Great Lakes world, it was the beaver itself that seemed to perpetuate problems for the Illinois: "The Beaver in Their district cause this Great carnage among us." Everything in Canada's policy seemed to be oriented around this animal, animated by "no other desire than to allow the vein of Beaver skins to flow." As for the Illinois, however, they knew their economic interests were completely different: "I admit that we do not kill as many Beavers as the People of the lakes," as one of the Illinois chiefs put it, "but our traders here are not Interested [in beavers], and do not supply our Enemies either with powder or with guns to kill us."[40]

To say that the beaver was the problem—that the animal itself "caused the great carnage"—is a perfect ecotone expression. It acknowledges what was an obvious fact: two different economies divided the region and incentivized new and unfamiliar alliances and rivalries among its peoples. Here was the heart of the matter. While it was true that the Illinois like Anakapita resented the "bad faith of the renards and of those who seek their Beaver-skins," it was also true that their northern neighbors had long resented the Illinois' efforts to build power through bison, as they now did in the west.[41] No doubt the Illinois must have been conscious that their own trade animal was—just as much as the beaver—the source of "great carnage." Complaining disingenuously to le Chat Blanc in 1725 of the way the beaver was causing their problems, the Illinois surely sensed how their own enslavement of neighbors, including captured Foxes—so deeply connected to the bison economy—was perhaps the bigger cause of their recent troubles. As they listed off the names of their recent losses to the Foxes, the Illinois conspicuously did not list their Fox victims.

Neither did the French at Illinois, who claimed not to be sure whether the Illinois still held onto Fox slaves but assured the Canadians that in any case their sympathies were with the Illinois. This was a calculated maneuver, and an important reality. Although "not interested" in beavers, the Illinois' French supporters at Fort de Chartres were very interested in bison. Recall that du Tisné himself had in 1719 made a trip to the Missouri Country, where he engaged the Osages and the Pawnees in diplomacy. Here was the future of the Illinois Country, and the future of the bison people. It was exactly what Checagou was implying to the King at Versailles, even at the very moment that Ouábessébau was conducting his diplomacy.[42]

Although surely not the truths that Le Chat Blanc's fact-finding expedition was hoping to find, all of these were indeed important realities. Above all, the Illinois' strategic commitment to division and not to peace among the lakes tribes was clear for all to see, as was their fundamental identity as bison people. So, too, was the essential reality of the Illinois' continuing possession of Fox slaves. Le Chat Blanc failed to "withdraw" those slaves, but as became clear in the following years, they were still there. Now Canadian officials, hearing du Tisné's and others' explanations of the situation, understood that the Illinois might never give them up. If that proved to be the case, then the French would just have to accept it. Diplomacy hadn't worked; "Gentle means" had failed.[43] In 1725 the Foxes renewed their attacks on the Illinois and the Illinois retaliated. Once again the French had to rethink their strategy.

◆ ◆ ◆

From the period around 1730, a large number of documents from the French perspective survive in French archives. The sheer quantity of documents, combined with their often vivid descriptions, has encouraged historians to put a great emphasis on the French intentions and understandings during the climactic period of conflict.[44] Yet what happened in 1730 was not only the result of French strategy but also the logical conclusion of Indigenous actors seeking advantage in a competitive borderland. Even at this late date, numerous observers understood that the Illinois could end war altogether by sending back the Fox captives, clearing the path for unity. But not only did that never happen, despite French urging, indeed the French soon were overcoming their own rivalries, coordinating Louisiana and Canadian agendas, in order to unite against the Foxes.

The end of the French hopes for peace came slowly. In 1726 Constant Le Marchand de Lignery looked to negotiate a new peace between the Foxes and the Illinois, promising to create a trading post at the Fox village of

Ouestatimong if they would put aside their hostility against the Illinois. Fox elder Ouchala assured Lignery that he could convince his people to relent. All he needed was for the Illinois to return the Fox slaves. When Ouchala died in 1727, however, his promise of peace died too. The Illinois sent back no prisoners. As they had several times previously, the French then contemplated turning decisively against the Foxes; Louisiana and Canada could no longer be divided, wedged apart by the beaver and the bison peoples. As one wrote, "It will be necessary to take measures conjointly with the nations of the two Colonies [Louisiana and Canada], to subdue [the Foxes] by force of arms."[45] When the French advanced a new policy of extermination, expressing their intention in terms that seem downright genocidal, they were not truly following Indian desires. But the basic idea of excluding the Foxes was certainly an Illinois priority.

The plan for an attack against the Foxes in 1728 was an echo of the 1715 plan, a vision of bringing together armies from Michilimackinac, Chicago, and other parts of the pays d'en haut to march on Ouestatimong, the large Fox village at the end of the Wisconsin Valley.[46] Just like in 1715, the Illinois importantly did not make the trip. Months before the rendezvous was meant to happen, Pierre-Charles de Liette accompanied an Illinois army to Chicago to conduct an initial raid on an outlying Fox village and prepare for the expedition to the north. Encountering a small party of Foxes while on their way, the Illinois staged a small but successful attack and then once again decided to sit out the more ambitious expedition into Fox territory. They were not present months later when an army of 450 French and 800 Indigenous allies left from Michilimackinac under the command of de Lignery. Perhaps this was fateful. The Foxes learned of the impending attack early enough to escape, as the French and allies burned their village and cornfields.[47]

After 1728, however, exclusion of the Foxes was surely underway. With their allies dwindling, the Foxes' position was growing weaker. "You may rest assured that that wicked nation can live no longer," one French official told the Kickapoos. "The King wishes their death." The Foxes were turned away by the Ioways, abandoned by the Mascoutens, the Sauks, and the Kickapoos. Now they went back to Nicolas-Antoine Coulon de Villiers, commandant at Fort Saint Joseph in modern-day Michigan, to plead for peace. The New France governor, Beauharnois, wrote that he was in favor of clemency for the Meskwakis, saying that "peace would be desirable." Of course, this was the French view, not the view of many of France's Indian allies. Sensing the Meskwakis' isolation and aiming to control that trade route to the Sioux themselves, in the spring of 1730 a group of Ottawas, Chippewas, Menonimees, and Ho-Chunks launched a surprise attack on the Foxes. Then the same groups surprised a Fox party returning from a bison hunt, killing many and taking captives to distribute to neighboring tribes in Michigan and Wisconsin, inviting those tribes to join in the renewed campaign against the Foxes. By this point the Foxes had lost five hundred people in a year.[48]

Suffering, the Foxes decided to try their "emergency plan" to find refuge in the east among the Iroquois and the English. They set out in late spring 1730 and traveled through the Illinois Country with nearly their entire population of around two thousand people. Passing through the Illinois Valley near the former site of Grand Village, they encountered a group of Illinois hunting bison. The Foxes attacked, capturing seventeen Illinois hunters. Escaping members of the Illinois party traveled quickly to Cahokia to recruit an army against the vulnerable Fox travelers. Realizing their mistake and the danger to which they now were exposed, the Foxes sent a small delegation to Cahokia to negotiate, but the Illinois rejected the offer of peace. The Illinois sent runners to the Kickapoos and the Mascoutens, who were also hunting bison in the area around the Illinois Valley. Together the force soon numbered six hundred, and they moved to intercept the Fox party near the headwaters of what is now the Sangamon River. From the north another Algonquian army arrived, comprised of Sacs, Potawatomis, and the Illinois' kinsmen, the Miamis. The entire anti-Fox army soon numbered fourteen hundred.[49]

The site was a small dip in the prairie on the edge of a large moraine. A creek ran through this depression, providing a source of water during the hot afternoons of midsummer. As the large and beleaguered Fox party moved across the vast prairie, exhausted and harassed by insects, they worried that they could not defend themselves if they kept moving. Instead they dug in, creating substantial earthworks and a good and impressive defense. Quickly French and Indian detachments from Illinois and the upper country surrounded them.[50]

Here is not the place to recount all the details of the siege, which have been explored in wonderful detail by other historians.[51] The standoff lasted for several weeks and featured many dramatic turning points. The point to emphasize is that the siege was—at its start and at crucial moments throughout—an Illinois-led affair. During the days' long siege, there were several moments of parley and potential peace. None was successful. Far from being the result of French frustrations turned to genocidal intent, the failure of these armistices was in great measure a function of Illinois refusal and power. Importantly,

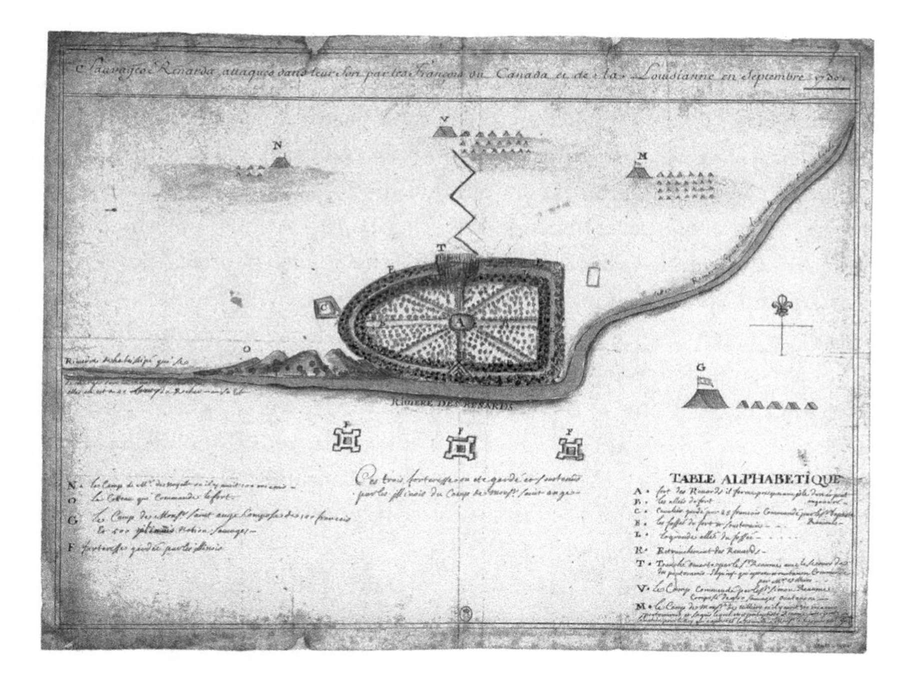

“Sauvages Renards Attaqués.” Map showing battlefield near modern-day Arrowsmith, Illinois. Anonymous Frenchman, 1730. Image courtesy Bibliothèque nationale de France.

in both cases the symbol of peace sent by the Foxes to call off the entire army was significant: a group of Illinois women captives.[52] Though de Villiers was in favor of peace in both instances, he knew he was not calling the shots, for “his party was not the most numerous.” As both the largest and the most influential army on the battlefield in 1730, the Illinois were also the most important and most determined, just as they had been in 1712 at Detroit. And, just as in that case eighteen years prior, it was the “Illinois who would not agree to any compromise.”[53]

The refusal to compromise was not the only element of that echoed events in Detroit in 1712. Another was the grim climax. Once again, although some escaped, the toll on the Foxes was massive. Writing in the spring of the following year, Louisiana governor Étienne de Périer wrote to the French minister with an estimate of casualties, announcing, “I have the honor to inform your Lordship of the defeat of the Foxes on the lands of Louisiana by the Illinois and the nations of the frontiers of Canada.” There were many uncertainties regarding the number of *slaves* taken on the battlefield, and disagreements over how many prisoners actually survived the maelstrom. But as Périer put it, the number of dead was massive: "The most definite thing that I have been able to learn from the Frenchmen who took part in this expedition is that eleven to twelve hundred Foxes, men, women and children, were killed."[54]

Just as in 1712, such numbers puzzled the French, who may have been no closer to a sympathetic understanding of the rationale behind "how much all these nations are irritated against the . . . Ottagamies." But we can understand both the symbolic and the material logic of this revenge. We know for whom the vengeance was taken, and, for instance, the litany of Illinois victims named specifically and for all to hear during le Chat Blanc's fact-finding. We know of the captives taken in 1722 during the Foxes' siege of the Peorias' village at Le Rocher. We know of previous Illinois losses from the more distant past—the captives taken by groups like the Iroquois and the Sioux, oft-times allies of the Foxes. To recompense for losses, and to build power, this warfare was needed.[55]

The logic of the war was also about something else. Just when this attack was underway, Checagou had recently returned from Paris where he had declared his strategic membership in the Missouri Valley world of trade in bison hides and plat côté (ribs), which represented a promising future for the Illinois and their neighbors. Departing for this trip with Bourgmont in 1724, the delegation had stowed away an important parcel: a collection of buffalo hides that they planned to personally gift to the king. Possibly painted, but almost certainly showpieces of what Charlevoix had once praised as the finest leather he had ever seen, these gifts were just the latest examples of bison hides brought to Paris to persuade the highest French authorities of their value. Like La Salle and Iberville, Checagou and his delegation were bringing evidence of the extraordinary commodity that could be extracted from the ecotone. As one of the chiefs put it, these were "several skins and works made by our wives."[56]

The words and speeches of this delegation were captured longhand by an anonymous scribe and preserved today in a curious collection of documents that now sits in the Bibliothèque Nationale in Paris. It is not clear who interpreted for the Indigenous speakers, nor how accurately the words were recorded. Evidently no one stopped to ask for more specifics about just who these wives, the hide makers, were. But as many eyewitnesses in Illinois villages in this period make clear, the answer would have been complex. Indeed, painting a different Illinois delegation traveling in Louisiana around this time, Alexandre de Batz made an important commentary on the violence of the Fox Wars when he foregrounded the Sauvagesse Renarde Esclave in his picture. Working a bison hide, then delivering tallow and plat côté, women like her remained a crucial factor in the bison economy that the Illinois and their neighbors looked forward to growing. This woman's labor—and the captive-raiding violence that helped produce it—were critical logics underlying what happened on the Illinois prairies in 1730.

◆ ◆ ◆

There is no doubt that the Fox Wars turned genocidal in the wake of 1730, as French officials began seriously contemplating what they had been discussing since 1712: the idea of eliminating the Foxes altogether. After the massive battle at Arrowsmith, the northern allies continued to pursue their advantage by attacking the remaining Foxes in several dramatic encounters. Most dramatically, the Hurons raised a party in 1732 against the Foxes, promising destruction. As Beauharnois made it clear, the purpose now seemed to be elimination. He wrote: "The [Indians] appear to me to be inclined to wipe out the [Fox] race, and I shall Maintain them in that disposition if the Renards fail to do what they promised me."[57]

There are important points to make here about this logic of elimination. First, if elimination were ever an Indigenous goal, the Indians pursued it in curiously ineffective ways. Indeed, Indigenous fighters in the ecotone had already chosen purposefully against eliminating the Meskwakis, which they clearly might have done at Arrowsmith, and they continued releasing Meskwaki prisoners taken on the battlefield, practicing the old ecotone gesture of "magnanimity." Between 1732 and 1735, as Beauharnois tried to organize several more expeditions against the Foxes to finally eradicate them, the Illinois and their Potawatomi neighbors conspicuously did not join.[58] Indeed, in the wake of 1730, the Illinois grew more suspicious of the French in general and hostile toward them. Rumors that their fate would soon resemble their Fox enemies—that the French would turn the same logic of exclusion and elimination against the Illinois—circulated in Kaskaskia and Cahokia.[59] As one Frenchman astutely put it, "You may imagine Monseigneur, that the [Indians] have their policy as we have Ours, and that they are not greatly pleased at seeing a nation destroyed for fear that their turn may come . . . We have had Recent proof of this among the Outawaois, who have begged for mercy for the Sakis, although they Had an interest in Avenging the death of their people and their great Chief."[60]

For the Illinois, it wasn't only that they feared they might be next; rather, it was that elimination of the Meskwakis had never been the goal of their fighting. Living in the ecotone, they positively benefited from having both more enemies and more friends. The Illinois did not thrive in an environment of unity and uniformity. Divisions, and the captives that embodied them, were at the heart of Illinois strategy. And while the French often exaggerated their own agency in bringing about and sustaining the Fox Wars, in fact the logic of elimination was the *one thing* that they likely could take credit for, a logic too easily confused in French writings as an Indigenous invention. Once they had accepted the animosity against the Foxes as a fact, the French were the ones that took it to an extreme and sought either all peace or *total war*. But when it came to this, the Indians of the ecotone, and the Illinois in particular, walked away. Violence and division, not elimination, was the goal.

After 1730 the Illinois retook their village near Starved Rock in the heart of contested bison-hunting territory.[61] They brought home and distributed a fresh round of captives. They sent some to the French as we will soon see. And they prepared for new conflicts: in 1736 they would begin fighting against the Chickasaws, their enemy to the southeast, once again. Most important, they went back to hunting. Sending out large bison hunts, they killed and processed numerous animals for trade. In 1735 they brought some of them to New Orleans, eager to remind the officials there of the bounty they could produce and the commodities they were prepared to trade. In making this reminder, as attested in de Batz's painting, the Illinois also brought with them an important resource in the production of these commodities, a Meskwaki slave. Likely recently captured at Arrowsmith, she could not have been a more important part of the recent history of the Illinois, whose special power had been built on bison and the energy streams it embodied, as well as captives and the enmities they embodied. Given its peculiar labor demands, the bison economy had introduced new incentives and material logics to old customs of captive-raiding and mourning war. It had accelerated Illinois aggression, especially when it came to dealing with their regional rivals, the Meskwakis. An important symbol and a material reality at the center of ecotone history, such slaves—and the bison they worked on—crucially shaped the violence of the Fox Wars.

CONCLUSION

COULIPA'S BODY AND THE POWER OF THE ECOTONE

IN THE LATE SUMMER OF 1732 A MESKWAKI MAN CALLED COULIPA (also written as "Goulipar") died alone in a prison inside the Hôtel de Cheusses in the city of Rochefort, France, some four thousand miles away from his homeland. Not much is known about his death besides a single notation in a registry. He may have received some care at a Rochefort hospital, where he was transferred four days before he finally expired on September 30. The scant documentary record of the event is part of the pathos of the whole situation. Wretched, isolated, distressed, and sick, Coulipa passed away, his imprisoned body a kind of symbol of what had happened to his people. Captured by the French. Reduced by violence. Overcome by the colonial encounter.[1]

Coulipa's body in a French jail symbolizes the domination of his people by the French, yet this is only part of the story. To see the rest we need to back up. We know that Coulipa's odyssey actually began months earlier and closer to home. Coulipa was initially captured as a prisoner of war during the climactic battle of the Fox Wars at present-day Arrowsmith, Illinois, in 1730. He was then sent to Governor Beauharnois in Quebec by Nicolas-Joseph de Noyelles, captain at Fort Saint Joseph and "leader" of one contingent of the French army that assembled to oppose the Foxes that summer. Coulipa's

Hôtel de Cheusses in Rochefort, France, where the imprisoned Meskwaki Coulipa spent some of his final days in 1732. © Musée national de la Marine/Michel le Coz.

journey to the French jail, and his submission to French power, began on the prairies of Illinois.

Importantly, moreover, Noyelles was just an intermediary. Coulipa's true captors were the Miamis and the Illinois. They were the ones who had led the battle at which Coulipa was captured. They were the ones who had refused to grant terms of peace to the Foxes, as Noyelles himself suggested. They were the ones who had given no quarter. By sending Coulipa to New France as a slave we know what they were doing: treating him as a symbol of enmity, wedging the Meskwakis away from any peace with the French and making the French accept their own priority of Fox isolation, "obliging them to become their enemies," as Raudot once put it. In this sense, if Coulipa's body in the French prison symbolizes the Foxes' victimization at the hands of colonists, it also symbolizes the power of the Illinois and the Miamis. In sending Coulipa to the French the Illinois speakers were using his body as a symbol in their own strategic power play.

The meaning of Coulipa's body becomes even more complex as we consider it more specifically within this Indigenous context. For just as we can imagine Coulipa's body in the French jail, we can imagine him in that Illinois camp on the prairie edge when his body was first captured several months before. Here is where his body first took on important meaning *to his captors* on the battlefield. We can imagine the scene as Coulipa was brought into the camp. Gathering together for this dramatic moment, the group probably inflicted some ritual violence on their captive, a cathartic act meant to foster community and release resentment on behalf of lost relatives. In this context the Illinois may have performed ceremony. They may even have given Coulipa some mataché, or body paint, to ready him for some sort of ritual of domination and conversion.

Indeed, mataché was undoubtedly a conspicuous part of the meaning of Coulipa's body in the Illinois camp. We know this because we know exactly what Coulipa's body looked like, thanks to an anonymous Quebec artist who made a portrait of Coulipa in 1731 as he waited to embark on the boat that would carry him across the Atlantic. As we can see, one of most notable things about his body in this painting is his mataché—face paint and a series of elaborate tattoos made across nearly his entire arms, chest, and legs.

Thinking about the specific circumstances of his capture can shape how we read the mataché on Coulipa's body. A possible—though perhaps unlikely—interpretation is that his mataché was actually placed there by the Illinois as a kind of "branding." Just as the Illinois frequently used mataché to incorporate outsiders, tattooing them as a kind of initiation, perhaps they tattooed Coulipa in this extravagant fashion as they welcomed him into the camp and prepared to send him on to Quebec. (*Nikikipenara*: *I mark my slave in order to recognize him*, as a term in the Miami-Illinois language puts it.[2]) A more likely interpretation, however, is that they are *his own* tattoos, placed on his body long before his capture in 1730. Indeed, perhaps he wore them as many ecotone men did, as a sign of specific military accomplishments and battlefield valor and the trophies of kills and prisoners taken. In this case Coulipa's mataché was part of what made him such a valuable captive in the Illinois camp; the Illinois likely took pride in subduing *this* particular captive's body, relishing their opportunity to subdue this highly distinguished—and decorated—person. Either way, it is conspicuous how nearly identical Coulipa's mataché marks are to the Illinois' own mataché designs. Note the formal similarity to the Illinois tattoos on the body of the warriors that de Batz had painted in New Orleans in 1735. Observe the zigzags down the torso, the bold dividing lines along the sternum and stomach, the rings on the shoulder. But also note the similarity between Coulipa's tattoo designs and

"Guerrier Renard"—Coulipa (1731). Courtesy Bibliothèque nationale de France.

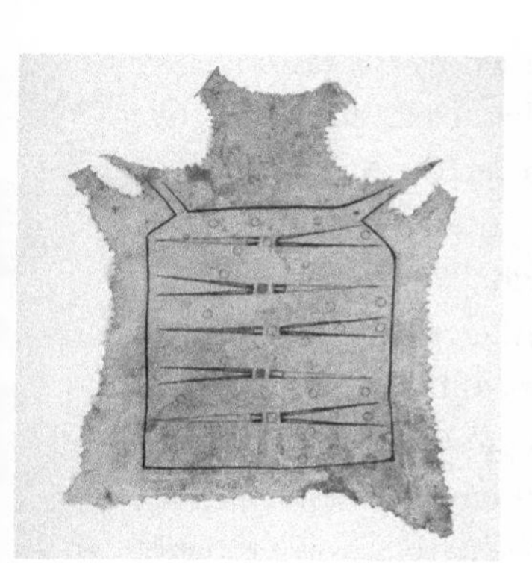

(left) Painted hide robe identified with the Illinois, eighteenth century. Courtesy Musée du quai Branly–Jacques Chirac. No. 71.1878.32.132. © RMN-Grand Palais/Art Resource, NY. *(center)* Illinois mataché. *Desseins de Sauvages de Plusieurs Nations* (detail) by Alexandre de Batz, 1735. Gift of the Estate of Belle J. Bushnell, 1941. Peabody Museum of Archaeology and Ethnology, Harvard University, 41-72-10/20. *(right)* Coulipa's mataché (detail). Courtesy Bibliothèque nationale de France.

the arrow motifs on an eighteenth-century hide painting associated with the Illinois, which is preserved today in the Musée du Quai Branly in Paris.[3]

However we interpret them, these similarities—what we might call the shared *tallgrass aesthetic* expressed in paintings and tattoos of the time—suggest a larger context for understanding the symbolism that Coulipa's body carried in the ecotone and across the ocean. Indeed, it seems plausible—if, of course, speculative—to read Coulipa's body as an archive of specific Indigenous history going back to the arrival of Algonquians like the Illinois and the Meskwakis in the prairie ecotone in the 1600s. Bearing the exact same motifs in its mataché as his rivals' and neighbors', Coulipa's body tells the story of encounters and exchanges between those groups, their long history together. It hints at the nature of those encounters, including in the context of warfare, slavery, and captivity. In this way, in other words, his body is a record of the very Indigenous trajectories that shaped his violent fate, a record of the complex history that defined this borderland. And although his body wound up in a French prison in the maritime town of Rochefort, it nevertheless reflected the currents of prairie history and the power of the Indigenous Midwest that helped to carry him there.

If Coulipa's mataché suggests the Indigenous trajectories that so powerfully shaped his life, however, so does one final important detail about

Coulipa's captured body, perhaps its most conspicuous feature. The artist who made Coulipa's portrait was obviously impressed by his tattoos, but perhaps even more impressed by one other physical characteristic of his body. Look closely at Coulipa's *legs*—long, muscular, powerful. As the painter made clear, they are the legs of an athlete, a *runner*.[4] The inscription on the painting makes their significance explicit. "Fox Warrior," it says "Dreaded by all the Nations, for their valor and speed, able to make 25 to 30 leagues per day without any other provision but plants and leaves of trees."

In today's terms twenty-five to thirty *lieus* (*leagues*) might amount to sixty to seventy-two miles. And while we might doubt this inscription's report about the source of Coulipa's caloric intake, the painter was probably not exaggerating about Coulipa's running range. Instead, as testified by countless eyewitness reports in colonial sources, here was a defining physical characteristic of the prairie people like Coulipa, the defining power of their bodies. If Coulipa could run for hours, he shared this ability with all of his people, and indeed many others in the prairie ecotone region. "They are the best runners in America," wrote Henri de Tonty of the Illinois. The Myaamias were "real and true greyhounds." People of the tallgrass ecotone like the Meskwakis and the Illinois were "the swiftest footed people *in the world*."[5] This was not just a toss-off description, but an earnest fact.

Far from a trivial detail, it was an important fact. This characteristic of Coulipa's body—his running prowess—came from somewhere. And while it is of course possible to understand it as just a cultural preoccupation, or even a genetic gift, the best way to understand Coulipa's running skill is as something more. For Coulipa's people and their neighbors, situated in their special environment, running was part and parcel of their relationship to it—the key to their locomotion across the edgy prairie and the shoreline of grass where they made their lives. Running was the energy that they spent to master this environment: to outrun fire, to run after game, to run in battle.

Most important, beginning sometime in the 1600s running was at the center of their partnership with an animal, their ungulate partner in this diverse grassland edge, the bison. Coulipa and his people—along with their neighbors—literally chased down bison across the grasslands in large teams on foot. Combining the work of their own bodies with the work of these giant grazers, the prairie people found a way to exploit the massive energy system in the C4 grasses and forbs of the tallgrass prairies. They fitted their bodies to the key physical and behavioral strategies of the *Bison bison*, the adaptive strategies of herding and running away as a response to predators. This practice—this work—shaped these prairie people, as this evocative painting of Coulipa attests, right down to their muscular legs.

Of course the same energy source also shaped much else about their lives besides their physical bodies. It fundamentally shaped their history. For instance, in the 1600s, as the prairie people rose to power along with their bison partners, they adopted large villages and cooperation with neighbors to enhance and sustain what was a volatile resource. Shaping a distinctive intergroup harmony, bison hunting also shaped a certain social order within villages, including especially a distinctive division of labor between men and women. With work organized into a balanced and complementary division of labor, villages like Coulipa's surely subscribed to a kind of communalism that was structured in part by the peculiar labor and the skills that pedestrian bison hunting demanded.

This balance was delicate, especially as it entangled with changing economic and demographic conditions at the moment of contact. As prairie people competed with neighbors in a complex borderland, groups like the Illinois increased their exploitation of bison to support large villages and build their power. In this context their small-scale captive raids turned more deadly. Now came a feedback loop, as the typical imperatives of kinship replacement that fueled warfare for many Indigenous North American groups entangled with the material demands of bison production. Villages tried first to maximize calories to support large populations and then to produce commodities for trade. Merging the logic of mourning war with an instrumental logic of slavery, these communities—particularly of the Illinois—relied increasingly on degraded captive labor to underwrite their maximization, allowing polygamous households to preserve the communalism and gender complementarity at the heart of their economy while still exploiting opportunities. In short, this human-bison assemblage shaped not just the prairie peoples' running ability; it also helped shape the slave-based militarism that defined their violent history in the contact era.

Both cause and effect of their relationship to bison, this militarism was extraordinary. For while prairie people like Coulipa were accomplished fighters and bison hunters, the important reality was that they were not alone. Just like Coulipa's fellow Meskwakis, Illinois men spent their time tallying prisoners captured and then marking them as distinct trophies on their skin. Disparaged by the French as lazy "gentlemen," in fact they spent time resting their powerful legs and preparing for the next bison hunt. The Illinois and the Foxes shared the same ambitious lifeway and viewed each other warily. And here is a key context for understanding the violence of this region in these years: just as they wore nearly the exact same tattoos, lived in the same environment, and practiced the same lifeway, they also shared a deep and contested history that went back to their migration to the region generations before.

Of course, beginning in the mid-1600s they also shared the disruption of colonialism and the ways in which exogenous forces like disease and guns intensified the ongoing trajectories of Indian power-building. And if bison and human history co-constituted one another in the ecotone, it would seem that colonialism and Indigenous trajectories likely did as well. Indigenous trajectories and multispecies entanglements determined *how colonialism came to matter.* If we consider Coulipa's body as a primary source document, surely one of the stories it tells powerfully is the way that colonization literally captured Indigenous lives and trapped them in new economic and political structures like capitalism and empire. Coulipa's sad fate in France—his solitary death in a prison while supposedly awaiting work as a galley slave—can be read rightly as a manifestation of these dynamics. But the Illinois are the ones who sent him to that prison too, and for particular reasons that grew out of continuity with the Indigenous past as much as the rupture of colonialism. To understand this fully we need to pay attention to Coulipa's powerful legs and his extraordinary tattoos, and recognize the story that they tell. The story of Coulipa's body is, like the 1730 battle in which he was captured, a story of both colonial and Indigenous power.

It is also—and in some ways just as importantly—a story of a more abstract power: the power of place, the power of a borderland, the power of the ecotone. Indeed, as this book has tried to show, this place was important. A main achievement of a recent generation of historians has been to "recenter" North American history around places and people often ignored by teleological, state-centered, national, and "frontier" narratives. Historians studying borderlands have done this perhaps most of all, focusing new attention on the interesting and dynamic locations and people in-between states or empires. In doing so they have recovered and highlighted the agency of the local people who subverted the hegemonic ambitions of competing states to create fugitive geographic and social configurations and exercise unexpected power and autonomy at the edge. In such places individuals and communities played imperial powers against each other to realize opportunities and advantages, often purposefully ensnaring themselves and building power in the middle.[6]

As I have argued, however, it was not only states and empires that created zones of division full of opportunity and unexpected power in the premodern past. In the prairie Midwest, ecology and related sociocultural divisions between largely *stateless* peoples created a different kind of borderland, but one no less full of history-shaping potential. Indeed, though emerging modern states were certainly present in this history, the people of the ecotone lived in a world formed just as profoundly by the politics of older chiefdoms—like Cahokia—and the consequences of their dissolution. And if that ancient Indigenous political world cast shadows over what was happening in the premodern prairies, it would seem that this Indigenous world was shaped even more profoundly by forces that had no coherent political center at all. In the prairie Midwest it was this special biogeography—the transition linking ecological and cultural zones—that allowed its inhabitants to take advantage of the diverse biomes and social world, to do the ecological equivalent of playing one against the other. States and chiefdoms had little to do with this. Rather, it was the tension in the tallgrass mosaic's important divide, as well as the Algonquian/Siouan-Caddoan division and the unsung agency of the bison, that created the opportunities and the advantages of this borderland. And while French eyewitnesses underestimated them, the currents of ecotone history were driven especially by the inhabitants' attempts to maximize the many advantages—and substantial power—that they could exploit in this heterogenous, unstable, and edgy landscape.

◆ ◆ ◆

In the 1730s, however, that power was changing, the world becoming less familiar. Some of what was changing was a matter of intensifying trajectories we already know about, particularly in the interaction of exogenous colonial forces with Indigenous patterns. For instance, in 1732 a smallpox epidemic devastated Illinois and Myaamia villages, causing a new round of population loss.[7] Whether this epidemic was another result of newly imported slaves carrying disease across the borderlands, as may have happened in the Great Smallpox Outbreak of the turn of the century, we cannot know. What we do know is that the demographic consequences for the Miamis—and presumably the Illinois—were severe. In this period French observers in their biggest permanent villages around Kaskaskia and Cahokia, began noting the relatively rapid decrease in Illinois population size.

An equally large change in the ecotone was not just a continuation of old trajectories, but a more categorical rupture: the shift to settler colonialism.[8] Although a hundred or so farmers had already put down roots in new French settlements in Illinois Country by the time of Pierre de Boisbriant's arrival in 1718, most colonists in the region had thus far been sojourners. In the 1720s and 1730s this all changed, as an especially large migration of European newcomers arrived in the now well-established villages near the junction of the Mississippi and Kaskaskia Rivers. Farmers began exporting wheat from Illinois to New Orleans, launching what would become a robust colonial trade. To support the efforts at wheat production, these French farmers also imported enslaved Africans, whose numbers ballooned as a percentage of the small population centers on the Mississippi River, thus joining a large contingent of Indigenous slaves already held by French families (no doubt "wedges" from their Indigenous neighbors). With these new arrivals the whole triad of settler colonialism was present: settler, native, slave. To be sure, in no way were Indigenous peoples fully eliminated from this new social world being redefined by a hybrid of plantation labor and idiosyncratic French farming techniques. But it is conspicuous that this new economy did go hand in hand with Indian decline—the near eradication of the Foxes and the stark decline of Illinois populations. For those French who envisioned the future of these colonies in terms of agricultural export, Indians could be ignored and marginalized. "We do with them what we want today," said one French observer of the Illinois, noting their smaller population.[9] These changes accompanied Indigenous demographic losses in the 1730s that many French observers noted.

In this context, and despite a "victory" at Arrowsmith in 1730, the Illinois struggled to regain their power. Making an important diplomatic visit to Paris in 1725, Illinois and Missouria chiefs had brought gifts of bison hides, hoping for a reestablishment of the long-promised tannery.[10] But having joined a bloc of powerful groups in the Missouri Valley, including the Osages and the Missourias, the Illinois now had to confront a pressing ecological reality of their own ecotone environment. North American bison indeed was about to have its day as a source of great energy and power for certain Indigenous groups. But as the Illinois had to realize, this was especially true for bison people located further west, in the heart of the shortgrass plains where bison were so plentiful that one French traveler in 1735 noted he had to wait three days to let an enormous herd pass.[11] As we have seen, the prairies may have been where French travelers *first* saw the bison, but by the mid-eighteenth century they well understood that it was nowhere near the best, nor the most stable, bison habitat of the continent. Continuing efforts to make the bison economy work in the Mississippi Valley failed in part because animal populations fluctuated so greatly, even as equestrian hunters of the west used their new partnership with horses to compensate for the great mobility of the herds.[12] All of this was to the detriment of the Illinois' prospects. The Wabash tannery remained just a notation on an old map; the massive tannery for le buffle in Corbeil remained safe from Illinois competition.

Meanwhile, if the Illinois had to contend with the marginality of the ecotone as bison habitat relative to points further west, they also had to contend with a new animal environment within the ecotone itself. With the arrival of settler colonists came different animals, particularly cattle, pigs, and horses. And while pigs were a nuisance that trampled their maize fields (*miincipi*), and while horses would eventually revolutionize the power dynamics for people on the shortgrass plains, cattle brought an immediate and local set of changes to the eastern prairie environment, even in the eighteenth century. That's because, like bison, cattle were an ideal partner to channel the grassland energy system of the transitional prairie. They occupied the same niche as bison, but did so in a very different way. Herders like bison, cattle were descended from the same long-ago ruminant bovine ancestors, possessing the ability to extract energy from the cellulose-rich grasses like big bluestem. But, unlike bison, they had become not just commensal, but full *servant livestock*. French settler colonists imported hundreds of them and their herds eventually grew fairly massive.

The difference between bison and cattle was in some ways subtle and not a categorical distinction between fully "domesticated" and fully "wild" animals. As we have seen, the prairie in the seventeenth century was an anthropogenic creation and bison were always to some extent a kind of cultivated species. Meanwhile, settler farmers often allowed their cattle herds to roam on the prairies in free-range fashion, thus the animals could adopt quite "wild" behavior.[13] Like bison hunting, cattle raising in the Mississippi Valley was entangled with systems of slavery, and it is likely that African slaves were part of how the settler colonial project managed meat and dairy operations. But for all its similarities, cattle husbandry was definitely more intensive; cattle produced more meat more predictably, especially as the settlers fenced the land. With these new herds of cattle, soldiers at Fort de Chartes and settlers in Louisiana could be supplied with meat produced by herdsmen in Kaskaskia, and they were.

Over time the new animal populations of the ecotone only increased. Indeed, it is important to note that the dominance of animal energy streams lasted a considerable period of time in the prairie Midwest, perhaps contrary to our usual assumptions. If we take for granted that this region—the modern corn belt—was made for row-crop agriculture, *nobody* could farm it for crops for a long time. The wheat economy of the American bottom grew substantial as early as the 1720s, but bottomland agriculture was an exceptional feat. In the prairie uplands—what would eventually become the true corn belt—the dense root structure of the tallgrass prairies made the sod much too dense to serve as cropland until John Deere's steel plows could turn the soil. Meanwhile, another problem in the uplands was moisture: the landscape was too wet for crops. In the town where I live, in the heart of the corn belt, the drainage projects that finally "solved" this problem were not accomplished until surprisingly late, at the end of the nineteenth century. The oldest building in our town testifies to this fact. "Cattle Bank," it curiously announces, looking like a building that belongs in Texas, a testament to the fact that early settlers—like Indigenous peoples before them—needed an animal intermediary to capitalize on the energy of the prairie landscape. Corn and soy did not replace cattle for many generations.

Cattle gradually did start to crowd out the bison, however. The wild grazers remained dominant in the Wabash and Illinois Valleys, a part of Illinois life for years; the last of the precontact bison were eradicated from Illinois around 1830. But bison would never be a source of power for Indigenous people of the eastern prairies after the mid-eighteenth century. Indeed, bison actually made an opportunity for colonial newcomers in a few notable episodes during the late-eighteenth century. For instance, occupying British armies took advantage of bison herds while traveling down the Wabash in the 1760s to seize the old French forts on the Mississippi in the wake of the Seven Years' War. In so doing, however, they treated the bison as a source of *found* energy and assumed its continuing stability, cutting into the bison population in a way that led to collapse. Encountering the first bison herds in the place where "you ordinarily begin to see the buffalo," the British and later-arriving settlers did not appreciate that in this particularly dynamic and unstable landscape of the eastern prairies, even supposedly "innumerable" herds were fragile.[14]

Meanwhile, Illinois power—not to mention Meskwaki power—relentlessly declined. Through the mid-eighteenth century the Illinois continued to rely on bison to preserve some opportunity in the face of change. In some years traders continued to come from New Orleans and Canada to buy plat côté in large quantities. The biggest market was always New Orleans, a city frequently short of food and struggling. As one observer put it in a *Relation* written in 1735, the export was sizable and "a great help to Lower Louisiana without which it would very frequently have the danger of starving." Sending this bison meat along with bison skins, the Illinois even adapted their mataché as a kind of trade commodity, promoting the skins as "bedcovers" that were "scraped, and painted by the Indians in a beautiful enough manner."[15] Decades later the famous early American traveler William Bartram observed how the Illinois had made something of a business of selling painted "Buffaloe hides," earning a reputation for these artworks among colonists and collectors in Europe.[16]

If these painted hides indeed resembled the painted hide pictured earlier, however, the repurposing of these objects for trade was a sign of how extensively the Illinois ecotone strategies were changing. Much evidence suggests that such painted hides, or *minohsaya* in the Miami-Illinois language, formerly were not merchandise but ceremonial objects used variously in diplomacy, the calumet ceremony, coming-of-age rituals, funerals, and perhaps especially in the context of adoptions. And if the commodification of mataché represented a major shift, meanwhile, another measure of profound change in the ecotone bison economy could be seen on annual hunts. Given the relatively small export market for bison meat, Illinois hunters in the mid-eighteenth century could satisfy the demand with minimal effort, leaving them bored with the event. This must be why French eyewitnesses began to observe hunting teams in the field chasing the bison purely for sport. A traveler in the Illinois Country around 1735, for instance, gives a powerful report of what was happening as an Illinois hunting party went to the Wabash prairies at that time. It is a stunning description, worth quoting at length. As the observer recalled: "To divert themselves [the Illinois went] to a prairie, which is found between the banks of the Wabash, where [they] found a lot of bison, which pass by. There are five or six [hunters] who each have their bows and arrows, which they shoot one after the other, into one of these bison. And when they don't have any more arrows, they run after the animal, taking the arrows out of the animal's flesh and then taking them out again a second time, and not quitting until [the animal] dies of loss of blood."[17]

Whew. Like the hunters themselves, we might need a moment to catch our breath after contemplating such a scene.

Consider where these hunters were: the north edge of the Wabash River Valley, a prairie in one of the flattest regions of what is now the United States. There was no cliff here, no buffalo "jump." These hunters were just running after the animal in an open field full of tallgrass. They were running so fast and in such a coordinated way that the six hunters could corral the animal and keep it from escaping. And if the whole act required amazing skill and know-how, consider the key part of the "sport" on display here: the way the hunters approached the massive wounded beast at full speed to retrieve spent arrows and then shoot them again.

It is extraordinary. Recall that many Frenchmen who confronted bison for the first time in the ecotone found them terrifying. And as city folks who heedlessly approach bison in parks and on farms today too frequently find out the hard way, this was a reasonable assessment of the animals' nature. Bison are wild and unpredictable tricksters, and they charge when wounded. So consider the skill, the speed, the energy, the confidence that this "diversion" involved. Consider the situated meaning that came from this extraordinary embodied, physical act.

It was at the heart of their lifeway. Cattle would soon dominate the prairies of the Midwest. Corn and soy would come next, massively simplifying the complex energy streams of the eastern prairie region and cutting out animal intermediaries altogether, or rather concentrating them in remote CAFOs (concentrated animal feeding operations), fully out of sight. Ironically, the very same factors that shaped the flux and contingency of the ecotone landscape in deep time were part of why latter-day settlers could so readily shape it into its modern uniform reality. But even as we appreciate the massive changes that they wrought, we need to acknowledge that the people who cultivated the industrial monocultural landscapes we know today were *not the first* to try to maximize energy streams of this region. Midwesterners, settler and Indigenous alike, should know the power of the people who came before, who mingled their energy with the landscape in a simply amazing way, by running after bison. Requiring extraordinary and peculiar kinds of *work,* this was the key to a lifeway and it was the key to an odyssey—an extraordinary early American rise and fall. The lifeway and its consequences are central to an understanding of the complicated history of Indigenous power in this special place, and they are key to understanding why the massive and violent changes of the colonial era played out here as they did.

◆ ◆ ◆

I've taken two trips to Arrowsmith, Illinois. The first time was in 2015, when I was lucky to be included in a group of friends who were hosting a visiting scholar at the University of Illinois. Setting out from Urbana early in the morning so as to return in time for the scholar's afternoon lecture, we drove west with the sunrise at our backs. It was early springtime and the landscape was still gray and brown, and no seedlings could yet be seen in the fields. Large pond-like puddles accumulated in many low areas, with water from recent rains that was stubbornly resisting the drainage infrastructure buried in the ground. I was oblivious to what this meant. What I saw on the roadside was just the natural shape of the rural landscape, flat and monotonous.

Our visitor knew where we were going; he was an expert on the history of the Fox Wars and had helped to confirm the location of the Fox fort during several archaeological studies years before. We had permission from the current owner of the farm to come and look from the barnyard, to gaze down into the heart of the site where the Fox fort had been situated. For my part I had prepared for the trip by quickly assembling a small cache of digital documents on my smartphone. Long a fan of visiting historical sites in person, I always liked to bring along evidence from the archive to compare against what we could see at the actual site. At Arrowsmith I wanted to fit what we saw in the landscape into what the old French maps depicted, using the documents to help me imagine what had happened in this place.[18]

Sure enough, it was easy to do, and powerful. When we got to the site, which is now a rare cow pasture in our region, our expert guide showed us approximately where the fort had been situated, known because archaeologists confirmed the presence of earthworks, metal, and other remains (including bison bones) from the weeks-long siege in the middle of the prairie.[19] A creek runs through the middle of the site, so it was pretty easy to match the various maps made by French engineers (such as the image on p. 193) to the physical landscape in front of me by lining up the curve of the creek. Soon I could see it. It was clear. I was seeing through the document and the landscape at once.

But as powerful and interesting as this was, I felt disappointed. Having studied early Illinois history for years, I had always been puzzled by the Fox Wars and the events of the summer of 1730 and hoped this site would give me clarity. Standing here—even with the French documents to help visualize the events—did not answer the biggest questions: What had really gone on here, and why? It was elusive. Walking back to the car I happened to scan the horizon. Perhaps I was *too stuck* to these sources, to the narrow view and the details of the old maps. I was being a good historian, but my perspective was limited.

I took a second trip to Arrowsmith in 2021. By this time, having lived in central Illinois for several more years, I had become a student of its natural history and fascinating, hidden ecology. New friends in Champaign in the ensuing years had taken me on several more early-morning field trips: to old rural cemeteries, "never-plowed" prairies, wetland restorations, the old-growth Trelease Woods at U of I, and—the newest kind of conservation efforts around here—prairie preserves like Nachusa and Midewin, with their reintroduced bison herds. Meanwhile, other skilled landscape interpreters, especially the painter Philip Juras, had helped me to envision ecological realities that I never previously managed to bring to life in my mind's eye simply by reading old accounts. With such experience under my belt, I returned to Arrowsmith with a newfound sensibility and a hunch. By this time, too, my phone was filled with other digital documents, including a second map—the first map in the introduction of this book.

Of course, in contrast to the detailed maps of the Fox fort found in the French archives, the map I brought was not the kind of map a historian usually would think of as a "primary source." Moreover, the sensibility and the spirit I now brought to the Arrowsmith site on my second trip was not the intense particularity that my discipline usually demands. But as I tramped around that cow pasture with the farm's owner and an archaeologist friend, I redid the exercise from years before. I envisioned where the fort had been, and I lined up the creek. This time my second map helped me to zoom out; I imagined the landscape stretching out toward the horizon.

The second map showed me something that should have perhaps been obvious all along but was seldom emphasized in all the histories I had read about this region and its premodern history: we were on the edge. This *particular* battle, so important in the history of the early Indigenous Midwest, took place on the blunt edge of the eastern prairie transition, in the heart of the former tallgrass mosaic. Looking out into the distance and back on the specific site in front of me, I could see that this was the important divider of the continent, not just in terms of ecology, but also of lifeway. Moreover, now I not only recognized the fact of this blurry divide, but I knew about its interesting history, including the fact of its relatively recent evolution, and the important ecological and human changes that defined it in the premodern past.

In this context the specifics of 1730 made much more intuitive sense as part of the ecotone's history, the outcome of long Indigenous trajectories. So often understood as the start of Indigenous decline—what came after—I now intuited that the Fox Wars make more sense in the context of a deep past—what came before. I now saw that these events were part of a long saga of Indigenous power, an older story growing right out of this special landscape just as surely as the few native grasses competing with the introduced alfalfa on this cow pasture under my feet. Contemplating those maps, I decided this was more than a hunch. I reminded myself that the whole melée of the summer of 1730 began when the Foxes ran into the Peorias while hunting bison right here in this prairie. Such a trigger, tied to the very texture of life in the ecotone borderland, is inextricably linked to that summer's crisis. And it is inextricably linked to a much bigger story, since that very same kind of encounter had surely happened countless times since 1600. It lay at the heart of the enmity between the Foxes and the Illinois.

The tallgrass ecotone landscape is not much noticed today. It is easy to pass through, or fly over, this landscape and completely miss it. Taking it for granted as "second nature," in William Cronon's words, even midwesterners hardly think about the massive landscape transformations that happened here or the radical ecological engineering that turned this massive division into a uniform monoculture. Related to this, as both cause and effect, we miss the stories that defined and were defined by this place, as Aldo Leopold said, before its transformation. But understanding the land and recovering the stories go hand in hand. We can recognize what was special about the region and its people in the past. Indeed, the French chronicler Antoine Raudot was partially correct: premodern history in the prairie region did come down to something essential about "the place where they lived." Surely that is one way to understand what Coulipa's body can tell us, the stories it illuminates. Living in this land we need to remember and resist its legacies of conquest and the ways in which colonialism captured Indigenous bodies and erased Indigenous history. But in doing that we need also to remember the power of Indigenous peoples and the way that power was shaped in relation to this forgotten place of division, this edge in the center of early America.

NOTES

INTRODUCTION

1 Risser, *The True Prairie Ecosystem*; R. Anderson, "The Eastern Prairie-Forest Transition"; Betz, *The Prairie of the Illinois Country,* chap. 1; Transeau, "The Prairie Peninsula."

2 R. Anderson, "Evolution and Origin ," 626; Wohl, *Wide Rivers Crossed*, 6.

3 Leopold, *A Sand County Almanac*, 124–26.

4 Cronon, *Nature's Metropolis*, xix, 56.

5 Prince, *Wetlands of the American Midwest.*

6 Cronon, "A Place for Stories."

7 For thoughts on parks and preservation in a somewhat analogous context, see Flores, *Caprock Canyonlands,* chap. 8.

8 Lauck, *The Lost Region*; and Hoganson, *The Heartland.*

9 Leopold, *A Sand County Almanac*, 127.

10 See O'Brien, *Firsting and Lasting*; and Buss, *Winning the West;* Brown and Kanouse, *Re-Collecting Black Hawk.*

11 Thwaites, *Jesuit Relations*, 59:105.

12 Bossu, *Travels*, 129.

13 DuVal, *The Native Ground.*

14 Richter, "War and Culture." Richard White calls the phenomenon of mourning war triggered by early colonial epidemics and fur markets in the Great Lakes an "engine of destruction" and offers this: "Never again in North America would Indians fight each other on this scale or with this ferocity." See White, *The Middle Ground*, 1. One question underlying the current study is why in the eastern ecotone the extreme violence that White associates with mid-seventeenth century Iroquois expeditions did not abate in 1701, at the end of the Beaver Wars, but continued throughout the Fox Wars.

15 Thwaites, *Jesuit Relations*, 59:145. Du Chesneau, "Memoir," 163; Rice, "War and Politics," 5.
16 Thwaites, *Jesuit Relations*, 58:67; Rushforth, *Bonds of Alliance*, 29; Perrot, "Memoir," 42–47, 260.
17 Richter, "War and Culture"; Rice, "War and Politics"; Holm, "American Indian Warfare."
18 R. White, *The Middle Ground*, 2.
19 For this common trope of Illinois decline see, among others, Hauser, "The Illinois Indian Tribe"; Shackelford, "The Illinois Indians," 15–16; Callendar, "Illinois," 678.
20 For a good example, see Hauser, "Warfare and the Illinois Indian Tribe." Even the best book on this period centers colonization as the key moving force to which Indians were usually reacting. See Edmunds and Peyser, *The Fox Wars*, especially xviii. An exception that has influenced my work strongly is Rushforth, "Slavery, the Fox Wars, and the Limits of Alliance."
21 Richter, *Facing East from Indian Country*.
22 Lee, *Masters of the Middle Waters*; Fenn, *Encounters*; Hämäläinen, *Lakota America*; Calloway, *One Vast Winter Count*. For an interesting discussion, see Ostler, *Surviving Genocide*, 7–8.
23 Hämäläinen, *The Comanche Empire*.
24 Zappia, *Traders and Raiders*.
25 Rushforth, *Bonds of Alliance*.
26 McDonnell, *Masters of Empire*.
27 Du Chesneau, "Memoir on the Western Indians, September 13, 1681," 162; Robert Cavelier La Salle, 1680, quoted in Margry, *Découvertes*, 2:96.
28 Edmunds and Peyser, *The Fox Wars*, 156.
29 Quoted in Calloway, *One Vast Winter Count*, 322–23.
30 Cadillac, "Memoir of Lamothe Cadillac," 67.
31 Lepore, *The Name of War*; Silver, *Our Savage Neighbors*.
32 For "humanity" at the heart of these stories, see Jacoby, *Shadows at Dawn*, 277–78; DeLay, *War of a Thousand Deserts*, 138; Farmer, "Borderlands of Brutality," 549–51.
33 Ostler and Shoemaker, "Settler Colonialism."
34 For dependency, see Hauser, "The Illinois Indian Tribe." My approach is strongly influenced by S. Warren, *The Worlds the Shawnees Made*.
35 My approach is influenced by Barr, "There's No Such Thing."
36 Rice, "War and Politics," 32.
37 On Indigenous "place" in early America, see Delucia, *Memory Lands*; L. Brooks, "Awikhigawôgan Ta Pildowi Ôjmowôgan"; and Thrush, *Native Seattle*. For Native space in the Midwest, see Sutterfield, "Aciipihkahki"; Governanti, "The Myaamia Mapping Project"; and McCoy et al., *Asiihkiwi Neehi*. This book takes a different "bioregional" approach to place.
38 See Flores, "Place"; and Flores, *Caprock Canyonlands*. See also Binnema, *Common and Contested Ground*; Hodge, *Ecology and Ethnogenesis*; and Zappia, "Indigenous Borderlands."

39 J. Brown, "The Prairie Peninsula."
40 Leopold, *Game Management*, 130–32, emphases in original. See Cronon, "Why Edge Effects?."
41 Clements, *Research Methods in Ecology*, 277; Rhoades, "The Ecotone Concept."
42 For the human "reshuffling," see Richter, *Before the Revolution*. For bison, a long discussion follows in chap. 2.
43 Andrews, *Killing for Coal*, 18; C. Jones, "Petromyopia," 36; Demuth, *Floating Coast*.
44 Smil, *Energies*, chap. 2. See also Isenberg, "Seas of Grass," 137–38.
45 Hämäläinen, "The Politics of Grass," 180.
46 LeCain, *The Matter of History*; Callon and Law, "Agency and the Hybrid Collectif"; De León, *The Land of Open Graves*.
47 See Russell, *Greyhound Nation*. For these other more familiar human-animal cultures, see Weisiger, *Dreaming of Sheep*; and LeCain, *The Matter of History*, chap. 4.
48 Certain theoretical insights from multispecies ethnography and new materialism have affinities with Indigenous cosmology. In attempting to bridge these ways of knowing by telling a material history, I acknowledge the limits of my own insight. See Todd, "An Indigenous Feminist's Take." See also Bennett, *Vibrant Matter*, xviii.
49 This formulation is borrowed from Callison, *How Climate Change Comes to Matter*.
50 Johnson, "On Agency"; O'Gorman and Gaynor, "More-Than-Human Histories."
51 Raudot, "Memoir," 383. For nearly identical "essentialist" thoughts about the Illinois, see Charlevoix, *History*, 5:130.
52 Raudot, "Memoir," 383.

1. SHORELINE OF GRASS

1 Jordan, "Between the Forest and the Prairie," 205; Madson, *Where the Sky Began*, 14; Betz, *The Prairie of the Illinois Country*, 1–7.
2 Cronon, *Nature's Metropolis*, 23, 25, 98.
3 Muñoz et al., "Defining the Spatial Patterns"; Hecht, "Domestication, Domesticated Landscapes, and Tropical Natures."
4 Pyne, *Fire*, 10.
5 Wedel, "The Central North American Grassland"; Sauer, "Grassland Climax, Fire, and Man"; Denevan, "The Pristine Myth." For examples of this kind of blunt interpretation, see Vale, *Fire, Native Peoples, and the Natural Landscape*; and Foreman, "The Myth of the Humanized Pre-Columbian Landscape." For a more subtle approach, see K. Anderson, *Tending the Wild*.
6 Cronon, *Changes in the Land*; Betz, *The Prairie of the Illinois Country*, 9.
7 Küchler, "Problems in Classifying," 515.
8 For "potential vegetation," see Küchler, "Potential Natural Vegetation," map of 1970. The prairie parkland transition is defined by Küchler as "mosaic of 66 and 91," or bluestem prairie and oak-hickory forest, respectively. For instability as a key feature of grasslands, see Isenberg, "Seas of Grass."

9 Thwaites, *Jesuit Relations*, 59:103, 60:155; Liette, "Memoir Concerning the Illinois Country," 306; Hennepin, *A New Discovery*, 1:149; Raudot, "Memoir," 384.
10 Thwaites, *Jesuit Relations*, 60:159, 58:97, 58:107.
11 Thwaites, *Jesuit Relations*, 66:269.
12 R. La Salle, *Relation of the Discoveries*, 87; Thwaites, *Jesuit Relations*, 59:103; Raudot, "Memoir," 384.
13 Thwaites, *Jesuit Relations*, 69:205.
14 Rodgers and Anderson, "Presettlement Vegetation."
15 Tobey, *Saving the Prairies*; Cronon, *Changes in the Land*; Worster, *Nature's Economy*; Winterhalder, "Concepts in Historical Ecology," 28–30; Barbour, "Ecological Fragmentation"; McCann, "Before 1492," 15.
16 Meine, "Foreword," xvii.
17 Cronon, "Paradigm Shift," 96.
18 Egan and Howell, "Introduction," 10–11.
19 Balée and Erickson, *Time and Complexity in Historical Ecology*. See also Crumley, *Historical Ecology*.
20 Wright, "Patterns of Holocene Climatic Change," 129.
21 J. Brown et al., "Toward a Metabolic Theory of Ecology."
22 Smil, *Energies*, 6; McNeill, *The Great Acceleration*, 7.
23 Clements, quoted in Wedel, "The Central North American Grassland," 60.
24 Madson, *Where the Sky Began*, 35, 38, 45.
25 Braudel, *The Mediterranean and the Mediterranean World*, 1:25.
26 Robertson, Anderson, and Schwartz, "The Tallgrass Prairie Mosaic," 59; Sala et al., "Primary Production of the Central Grassland"; Gleason, "The Vegetational History of the Middle West."
27 Wedel, "The Central North American Grassland," 41–42; Anderson, "Evolution and Origin," 628–31; Smil, *Energies*, 43.
28 King, "Late Quaternary Vegetational History," 45. See also Borchert, "The Climate of the Central North American Grassland," 19–25; Changnon et al., "Climate Factors"; Madson, *Where the Sky Began*, 34; Wedel, "The Central North American Grassland," 45; and Knapp and Medina, "Success of C4 Photosynthesis," 253.
29 E. Kellogg, "Evolutionary History of the Grasses," 1201; Osborne and Beerling, "Nature's Green Revolution," 173; R. Anderson, "Evolution and Origin," 627, 633; Smil, *Energies*, 44–45.
30 Gleason, "The Vegetational History of the Middle West," 41.
31 Clements et al., quoted in Wedel, "The Central North American Grassland," 59.
32 Knapp and Medina, "Success of C4 Photosynthesis in the Field," 253.
33 Axelrod, "Rise of the Grassland Biome," 166; R. Anderson, "Evolution and Origin," 627.
34 King, "The Prairies of Illinois"; Thomson, *The Shaping of America's Heartland*, chap. 2; Kolata and Nimz, *Geology of Illinois*, chaps. 6–11.
35 King, "Late Quaternary Vegetational History," 45.

36 For "watery world" see Morgan, *Land of Big Rivers*; Wohl, *Wide Rivers Crossed*, 152, 156.
37 Smil, *Energies*, 42–43.
38 Risser, "The Status of the Science," 319; Gosz, "Ecological Functions in a Biome."
39 Sauer, "Grassland Climax, Fire, and Man," 20; Axelrod, "Rise of the Grassland Biome," 166.
40 Clements, quoted in Wedel, "The Central North American Grassland," 59.
41 R. Anderson, "Evolution and Origin," 635.
42 King, "The Prairies of Illinois," 3; King, "Late Quaternary Vegetational History," 57; R. Anderson, "Evolution and Origin," 631; Robertson, Anderson, and Schwartz, "The Tallgrass Prairie Mosaic," 59; Styles and McMillan, "Archaic Faunal Exploitation," 40–43; Webb et al., "Climatically Forced Vegetation Dynamics"; Wright, "The Dynamic Nature of Holocene Vegetation," 587–90; Nelson et al., "Response of C3 and C4 Plants."
43 Wright, "History of the Prairie Peninsula"; King, "Post-Pleistocene Vegetational Changes."
44 Grimm, "Chronology and Dynamics of Vegetation Change"; Baker et al., "Patterns of Holocene Environmental Change."
45 Boggess and Geis, "The Prairie Peninsula," 92; Robertson, Anderson, and Schwartz, "The Tallgrass Prairie Mosaic," 57.
46 Sauer, "Grassland Climax, Fire, and Man"; McClain et al., "Patterns of Anthropogenic Fire."
47 Pyne, *Fire*, 3.
48 Nelson et al., "The Influence of Aridity and Fire"; Nelson and Hu, "Patterns and Drivers."
49 R. Anderson, "The Historic Role of Fire"; Allen and Palmer, "Fire History"; Camill et al., "Late-Glacial and Holocene Climatic Effects"; Collins and Wallace, *Fire in North American Tallgrass Prairies*.
50 Andrews, *Coyote Valley*, 33.
51 J. Morris et al., "Holocene Fire Regimes," 1451–52; Pyne, *Fire*, 19.
52 R. Anderson, "Evolution and Origin," 628; Van Nest, "Late Quaternary Geology," 352–54.
53 Axelrod, "Rise of the Grassland Biome," 187.
54 R. Anderson, "Evolution and Origin," 635.
55 Sauer, "Grassland Climax, Fire, and Man," 20.
56 R. Anderson, "The Historic Role of Fire."
57 R. Anderson, "Evolution and Origin," 632.
58 Vinton et al., "Interactive Effects of Fire"; Allred et al., "Ungulate Preference for Burned Patches"; Knapp et al., "The Keystone Role of Bison"; Collins et al., "Modulation of Diversity."
59 R. Anderson, "Evolution and Origin," 638, 641.
60 Williams, Shuman, and Bartlein, "Rapid Responses of the Prairie-Forest Ecotone"; Hupy and Yansa, "Late Holocene Vegetation History."

61 Fei et al., "Divergence of Species Responses," 2. For instability on the western side of the tallgrass ecotone, see Wohl, *Wide Rivers Crossed*, 6.
62 Axelrod, "Rise of the Grassland Biome," 165.
63 Flores, *American Serengeti.*
64 McElrath and Emerson, "Concluding Thoughts," 845–47.
65 Struever, "Woodland Subsistence-Settlement Systems"; Binford, "Willow Smoke and Dogs' Tails."
66 Griffin, "Climatic Change"; B. Smith, *Rivers of Change*, 207.
67 Benson, Pauketat, and Cook, "Cahokia's Boom and Bust."

2. SPECIES SHIFT

1 Flores, *American Serengeti*, 124.
2 Cronon, Gitlin, and Miles, "Becoming West."
3 Crosby, *The Columbian Exchange*; Crosby, *Ecological Imperialism*; Cronon, *Changes in the Land*; Mann, *1493*; Brooke, "Ecology." For critiques, see Hämäläinen, "The Politics of Grass," 173–75; and L. Warren, "The Nature of Conquest."
4 Emerson and Lewis, *Cahokia and the Hinterlands.*
5 Ethridge, *From Chicaza to Chickasaw*, 18–24.
6 Emerson and Hedman, "The Dangers of Diversity"; Anderson and Smith, "Pre-Contact," 15; Kehoe, *North America before the European Invasions*, 140–45.
7 Mann, *1491*, 194; Pollan, *The Omnivore's Dilemma*, 23–24.
8 Rindos and Johannessen, "Human-Plant Interactions," 42–43.
9 Benson, Pauketat, and Cook, "Cahokia's Boom and Bust."
10 Benson, Pauketat, and Cook, "Cahokia's Boom and Bust," 474.
11 Muñoz et al., "Cahokia's Emergence and Decline"; Lopinot and Woods, "Wood Overexploitation." See also A. White et al., "Fecal Stanols," 5462; and Benson, Pauketat, and Cook, "Cahokia's Boom and Bust," 476–78.
12 Hämäläinen, "The Changing Histories of North America"; Kehoe, "Cahokia, the Great City."
13 Mann, *1491*, 266–67; Diamond, *Collapse.*
14 A. White et al., "After Cahokia," 2.
15 Lee, *Masters of the Middle Waters.* For general reframing of collapse narratives toward concepts like "transition," see Faulseit et al., "Collapse, Resilience, and Transformation."
16 Widga, "Niche Variability in Late Holocene Bison"; Widga, "Bison, Bogs, and Big Bluestem."
17 Martin and Harn, "The Lonza-Caterpillar Site"; Harn and Martin, "Early Confrontations.'"
18 Parmalee, "The Faunal Complex of the Fisher Site"; McMillan, "Perspectives on the Biogeography," 76–84, 99.
19 The most authoritative summary of the recent research is McMillan, "Perspectives on the Biogeography." See also Widga and White, "An Ecological History"; and J. White, "A Review of the American Bison."

20 LeCain, *The Matter of History*, 147; Lott, *American Bison*, 48; Smil, *Energies*, 44–45.
21 Lott, *American Bison*, 49, 64; Demuth, *Floating Coast*, 16.
22 Flores, *American Serengeti*, 20–21, 113; Lott, *American Bison*, 62–65, 97; Flores, *Caprock Canyonlands*, 16.
23 Lott, *American Bison*, 65.
24 Speth, "Communal Bison Hunting," 280; Hamilton, "The Genetical Evolution."
25 LeCain, *The Matter of History*, 151.
26 Bamforth, *Ecology and Human Organization*; Binnema, *Common and Contested Ground*, 18–19.
27 Lott, *American Bison*, 21–22.
28 Oliver, *Ecology and Cultural Continuity*; Arthur, *An Introduction*. For a brilliant analysis of seasonality and its impact on Northern Plains history, see Binnema, *Common and Contested Ground*, chap. 2.
29 Dillehay, "Late Quaternary Bison Population Changes." See also Huebner, "Late Prehistoric Bison Populations."
30 Wallace et al., "Scale of Heterogeneity."
31 Flores, *American Serengeti*, 124; Flores, "Bison Ecology." See also Cooper, "Bison Hunting"; and Boehm, "Were Bison Predictable Prey?."
32 Shay, "Late Prehistoric Bison and Deer Use," 199–200.
33 Widga and White, "An Ecological History"; Widga, "Niche Variability."
34 For the rapid response of tallgrass prairies to drought conditions, which would have produced better mixes of grama grasses over big bluestem, see Shay, "Late Prehistoric Bison and Deer Use," 199.
35 Flores, *American Serengeti*, 23–24, 124–25.
36 Shay, "Late Prehistoric Bison and Deer Use," 197, 201; Tankersley and Lyle, "Holocene Faunal Procurement."
37 Cobb and Butler, "The Vacant Quarter Revisited"; Schroeder, "Current Research."
38 A. White et al., "After Cahokia," 2; J. Brown and Sasso, "Prelude to History," 205.
39 Cobb and Butler, "The Vacant Quarter Revisited," 625; Kehoe, *North America before the European Invasions*.
40 For the idea of an interaction zone, see J. Brown, "The Prairie Peninsula."
41 Emerson and Brown, "The Late Prehistory," 86, 94; Esarey and Conrad, "The Bold Counselor Phase"; Griffin, "A Hypothesis"; Green and Benn, *Oneota Archaeology*; Henning, "The Oneota Tradition."
42 See Ritterbush, "Drawn by the Bison"; J. Brown, "What Kind of Economy?"; Michalik, "An Ecological Perspective." For "broad spectrum," see Cooter, *Ecotones and Broad Spectrum Economies*.
43 J. Brown and Sasso, "Prelude to History," 206.
44 For Mississippians' narrow dependence on maize and its health effects, see Goodman and Armelagos, "Disease and Death."
45 Faulkner, *The Late Prehistoric Occupation*. See also Benn, "Hawks, Serpents, and Bird-Men"; and Gibbon, "Cultural Dynamics."
46 Schroeder, "Current Research," 316–18.

47 Boszhardt and McCarthy, "Oneota End Scrapers"; J. Brown, *Aboriginal Cultural Adaptations*; Emerson and Brown, "The Late Prehistory," 86–87.

48 Michalik, "An Ecological Perspective"; Tankersley and Lyle, "Holocene Faunal Procurement."

49 Brown and Sasso, "Prelude to History," 205; Schroeder, "Current Research," 336; Sasso, "La Crosse Region Oneota Adaptations"; McMillan, "Perspectives on the Biogeography," 107.

50 McMillan, "Perspectives on the Biogeography," 81; Bluhm and Liss, "The Anker Site," 125; Robert L. Hall quoted in McMillan, "Perspectives on the Biogeography," 80.

51 Emerson and Brown, "The Late Prehistory," 82. For the theory of Danner pottery as interlopers at Fort Ancient sites, see Mazrim, "The Danner Series Pottery," 29–30.

52 Emerson and Brown, "The Late Prehistory," 90; Mazrim and Esarey, "Rethinking the Dawn of History."

53 Emerson and Brown, "The Late Prehistory," 89; Tankersley, "Bison and Subsistence Change."

54 Jakle, "The American Bison," 301–2.

55 Griffin and Wray, "Bison in Illinois Archaeology," 24–25; Boszhardt, "Turquoise, Rasps and Heartlines," 364; Henning, "Cultural Adaptations," 193–94; McMillan, "Bison in Missouri Archaeology," 87; Sasso, "La Crosse Region Oneota Adaptations," 343–45; J. Brown and Sasso, "Prelude to History," 210.

56 Flores, *American Serengeti*, 125.

57 S. White, *A Cold Welcome*; Flores, *American Serengeti*, 24–25, 125–26; McMillan, "Perspectives on the Biogeography," 69.

58 See Griffin and Wray, "Bison in Illinois Archaeology," 25.

59 See Stewart, *Forgotten Fires*, 120–21; Gleason, "The Vegetational History"; and Gleason, "The Relation of Forest Distribution."

60 For discussion, see A. White et al., "After Cahokia," 7. See also Muñoz, "Forests, Fields, and Floods," 58.

61 The rising concentration of charcoal at year 1500 is subtle in general, but when examining cellular charcoal produced from partially combusted grasses it is dramatic. See Muñoz, "Forests, Fields, and Floods," 61.

62 Van Nest, "Late Quaternary Geology," 351–52. This inference is also supported by the research of McClain and colleagues, who conclude anthropogenic fires were far more common than lightning-set fires over a long stretch of early historical accounts. McClain et al., "Patterns of Anthropogenic Fire."

63 Muñoz et al., "Defining the Spatial Patterns." For speculation on the relationship beteen human activity and the presence of bison east of the Mississippi, see J. White, "A Review of the American Bison," 5–6.

64 Stewart, "Burning and Natural Vegetation"; Stewart, *Forgotten Fires*, 117–26. The authoritative study of eastern tallgrass prairie burning from historical accounts is McClain et al., "Patterns of Anthropogenic Fire." See also Betz, *Prairie of the Illinois Country*, 38–40, and chap. 4.

65 R. La Salle, *Relation of the Discoveries*, 220.

66 For an excellent discussion, see Hecht, "Domestication, Domesticated Landscapes."
67 Leopold, *Game Management*, 21.
68 LeCain, *The Matter of History*, 150.
69 O'Gorman and Gaynor, "More-Than-Human Histories," 717; Russell, "Fauna."
70 J. Brown and Sasso, "Prelude to History," 212–15.
71 Hämäläinen, "The Rise and Fall."
72 Greer et al., "Land Cover of Illinois in the Early 1800s."

3. THE RUN-UP

1 De Tonti, *Relation of Henri de Tonty*, 29; R. La Salle, *Relation of the Discoveries*, 191; Liette, "Memoir," 381; D'Artaguiette, "Journal of Diron D'Artaguiette," 73.
2 See the illustration in the conclusion, "Guerrier Renard," courtesy Bibliothèque Nationale.
3 Sayre, *Les Sauvages Américains*; Chaplin, *Subject Matter*.
4 Raudot, "Memoir," 364, 388. For other testimonies, see Thwaites, *Jesuit Relations*, 66:229; Charlevoix, *Journal*, 2:193; and R. La Salle, *Relation of the Discoveries*, 257.
5 Nabokov, *Indian Running*.
6 See, for instance, Gilbert, *Hopi Runners*; and Dyreson, "The Foot Runners."
7 See, for instance, McDougall, *Born to Run*. See also Bramble and Lieberman, "Endurance Running."
8 Latour, *Reassembling the Social*; O'Gorman and Gaynor, "More-Than-Human Histories." On "hybridity," see Sutter, "The World with Us"; and LeCain, *The Matter of History*.
9 LeCain, *The Matter of History*, 142.
10 Callon and Law, "Agency and the Hybrid Collectif."
11 Barsh and Marlor, "Driving Bison and Blackfoot Science."
12 In examining the "the energetics of the bison drive" I have been inspired by Wilson and Davis, "Epilogue," 317–18.
13 The original of Liette's account is in the Newberry Library, Chicago (VAULT Ayer MS 293, vol. 3); this description is based on Liette, "Memoir," 307–14.
14 Thwaites, *Jesuit Relations*, 66:287; St. Cosme, "Voyage of St. Cosme," 350; de Tonti, *Relation of Henri de Tonty*, 29; Liette, "Memoir," 304.
15 Hennepin, *A New Discovery*, 2:150–52; Thwaites, *Jesuit Relations*, 67:167; Margry, *Découvertes*, 1:261.
16 Hennepin, *A New Discovery*, 1:145–50, 2:652–56; Perrot, "Memoir," 121–26; Raudot, "Memoir," 407–8; Charlevoix, *Journal*, 1:188–90; Thwaites, *Jesuit Relations*, 65:73–75; N. La Salle, *Relation of the Discovery*, 81–83.
17 Arthur, *An Introduction*.
18 Charlevoix, *Journal*, 1:188; Perrot, "Memoir," 121.
19 Liette, "Memoir," 304, 309; Thwaites, *Jesuit Relations*, 66:287.
20 Liette, "Memoir," 310.
21 Liette, "Memoir," 311.

22 See Bamforth, *Ecology and Human Organization*, 8–9, 22. See also Fenn, *Encounters*, 66–67.

23 On danger, see Hennepin, *A New Discovery*, 2:645; and Thwaites, *Jesuit Relations*, 59:111.

24 Others confirmed this number. Hennepin, *A New Discovery*, 1:147. Charlevoix noted that few parties ever killed fewer than fifteen hundred in a single outing. Charlevoix, *Journal*, 1:188.

25 Krech, *The Ecological Indian*, 132–33.

26 See also Hennepin, *A New Discovery*, 2:521.

27 Raudot, "Memoir," 408.

28 Although the literature about communal bison hunting on the plains is plentiful, few scholars have focused on the practice in the context of the tallgrass. For exceptions, see Shay, "Late Prehistoric Bison and Deer"; and Lehmer, "The Plains Bison Hunt."

29 Robbins, *Lawn People*.

30 See R. White, "'Are You an Environmentalist?.'"

31 R. White, *The Organic Machine*; R. White, "Animals and Enterprise." For the usefulness of thinking about nonhuman *workers* in history, see Russell, *Evolutionary History*, 137; and Sutter, "Nature's Agents?." See also Greene, *Horses at Work*.

32 Demuth, *Floating Coast*, 16.

33 For the idea of the "secondary food cycle," see R. White, *The Roots of Dependency*, 27–28, 98; and Isenberg, *The Destruction of the Bison*, 38–39.

34 Robert Betz performs a similar analysis, focusing on three main biozones in the prairie edge. See Betz, *The Prairie of the Illinois Country*, 41–44.

35 McCafferty, "Illinois Voices," 127. See also Gonella, "Myaamia Ethnobotany," 113–17.

36 Gonella, "Myaamia Ethnobotany," 19–20, 86.

37 Gonella, "Myaamia Ethnobotany," 20.

38 Liette, "Memoir," 340, 345–46; Perrot, "Memoir," 113–19; Gonella, "Myaamia Ethnobotany," 20, 31, 48. See also Morgan, *Land of Big Rivers*.

39 Liette, "Memoir," 320. Prairie peoples' location at the edge of both the forest *and* the prairie gave them access to game from both habitats, and sometimes forest game was just as bountiful as bison and elk. For instance, see Thwaites, *Jesuit Relations*, 60:149–51.

40 Gonella, "Myaamia Ethnobotany," 31.

41 Gonella, "Myaamia Ethnobotany," 19, 21; Liette, "Memoir," 347.

42 Scott, *Against the Grain*, 46, 51, 61.

43 Sassaman, "Complex Hunter-Gatherers"; see also Bamforth, *Ecology and Human Organization*.

44 Barsh and Marlor, "Driving Bison and Blackfoot Science."

45 Perrot, "Memoir," 119; Shay, "Late Prehistoric Bison," 200; Hennepin, *A New Discovery*, 2:563, 1:148–49.

46 For the more predictable seasonal patterns of the Northern Plains, see Binnema, *Common and Contested Ground*, chaps. 1–2. See also Vickers, "Seasonal Round

Problems," 55; and Hanson, "Bison Ecology," 93–113. For bison grazing patterns on the tallgrass, see Schuler et al., "Temporal-Spatial Distribution."

47 Thwaites, *Jesuit Relations*, 66:219, 255; Pinet, *Myaamia-Illinois Dictionary*, 109, ILDA.

48 Thwaites, *Jesuit Relations*, 58:97.

49 Thwaites, *Jesuit Relations*, 65:73, 66:253; Walthall, Norris, and Stafford, "Woman Chief's Village."

50 Hennepin, *A New Discovery*, 1:147, 2:519; Perrot, "Memoir," 120.

51 See Thwaites, *Jesuit Relations*, 55:195.

52 Isenberg, *The Destruction of the Bison*, 39–47.

53 Pianka, *Evolutionary Ecology*, 185–86; Shay, "Late Prehistoric Bison," 194–97; Flores, "Bison Ecology," 476–77.

54 Hennepin, *A New Discovery*, 1:242, 1:147, 2:563.

55 Hennepin, *A New Discovery*, 2:519–20.

56 For "irenansecana," see Pinet, *Myaamia-Illinois Dictionary*, 155, ILDA.

57 See also Thwaites, *Jesuit Relations*, 67:157–59.

58 Especially revealing of prairie peoples' ideas about animals was their approach to *domestic* animals. See Breen, "'The Elks Are Our Horses.'"

59 Thwaites, *Jesuit Relations*, 51:35.

60 See, for instance, text 22 in Costa and Baldwin, *Myaamia Neehi Peewaalia*, 96–97.

61 Thwaites, *Jesuit Relations*, 66:237.

62 Raudot, "Memoir," 392; Thwaites, *Jesuit Relations*, 59:221, 67:157.

63 D'Artaguiette, "Journal," 58.

64 For examples of trickster tales from prairie cultures, see Michelson, "Contributions to Fox Ethnology"; and Costa and Baldwin, *Myaamia Neehi Peewaalia*.

65 Hennepin, *A New Discovery*, 1:148, 1:280; Charlevoix, *Journal*, 1:188.

66 Charlevoix, *Journal*, 1:188; Liette, "Memoir," 312–13; Perrot, "Memoir," 120.

67 Hennepin, *A New Discovery*, 1:147; Perrot, "Memoir," 124.

68 Liette, "Memoir," 312–13.

69 Hennepin, *A New Discovery*, 1:147; Thwaites, *Jesuit Relations*, 65:73.

70 Smits, "The 'Squaw Drudge.'"

71 T. Edwards, *Osage Women and Empire*, 2–3; Isenberg, *The Destruction of the Bison*, 93–100.

72 Thwaites, *Jesuit Relations*, 66:251.

73 D'Artaguiette, "Journal," 73; Pratz, *The History of Louisiana*, 309.

74 Thwaites, *Jesuit Relations*, 66:231.

75 Skinner, "Traditions of the Iowa Indians"; Trudeau, *A Fur Trader on the Upper Missouri*, 301. See also linguistic clues in the Miami-Illinois language, such as words like *Memeti8sseta8i* 'Let us play who will run the best'." Largillier Gravier, 270, ILDA.

76 Liette, "Memoir," 377.

77 Skinner, "Traditions of the Iowa Indians," 435, 441, 457.

78 Michelson, "Contributions to Fox Ethnology," 1–50.

79 Thwaites, *Jesuit Relations*, 55:77; Lahontan, *New Voyages*, 204.

80 Sheehan, *Running and Being*; Swann et al., "Optimal Experiences in Exercise."
81 Vergunst and Ingold, *Ways of Walking*; Ingold, *The Perception of the Environment.*
82 In the episode that prompted this remark, Liette witnessed the Illinois chasing a group of does, running them down to such a point of exhaustion that they could literally put a hand on the back of each animal before killing it. Liette, "Memoir," 319–20.
83 Charlevoix, *Journal*, 2:193.
84 Hennepin, *A Description of Louisiana*, 282; Raudot, "Memoir," 388. See also Hennepin, *A New Discovery*, 2:488; Margry, *Découvertes*, 5:86, 493; and Sabrevois, in WHC 16:372.
85 Hennepin, *A New Discovery*, 1:147.
86 La Potherie, "History," 322; de Tonti, "Memoir on LaSalle's Discoveries," 303.
87 "Relation of Sieur de Lamothe Cadillac" (1718), in WHC 16:361.
88 Galloway, *Choctaw Genesis*; Merrell, *The Indians' New World.*
89 Hodge, *Ecology and Ethnogenesis.*
90 Shackelford, "The Illinois Indians," 22.
91 Brown and Sasso, "Prelude to History," 215; Gibbon, "Cultural Dynamics."
92 LeMenager, *Living Oil.*

4. EDGE AND WEDGE

1 R. La Salle, *Relation of the Discoveries*, 197, 215, Thwaites, *Jesuit Relations*, 62:71–73. See also Margry, *Découvertes*, 2:124–34.
2 For "engine of destruction," see R. White, *The Middle Ground*, 1. Daniel Richter has usefully written that the cyclical feedback loop of mourning war, disease, and captive-raiding was a "spiral" of violence in contact-era indigenous societies. See Richter, "War and Culture." See also Richter, *The Ordeal of the Longhouse*, 74; and Parmenter, *The Edge of the Woods*, 77–86.
3 L. Kellogg, *The French Régime in Wisconsin and the Northwest*. For the narrative of decline among the contact-era Illinois, see Blasingham, "The Illinois Indians"; and Hauser, "The Illinois Indian Tribe."
4 R. La Salle, *Relation of Discoveries*, 195, 255. See also Margry, *Découvertes*, 1:527–28.
5 R. White, *The Middle Ground*, 2. While White emphasizes an "imported imperial glue," this discussion foregrounds the ways in which captive exchange constituted an *Indigenous* glue for making alliance. Rushforth, *Bonds of Alliance*, 29.
6 R. White, *The Middle Ground*, 11; Richter, *The Ordeal of the Longhouse*, 50–74.
7 R. White, *The Middle Ground*, 15–19; DeMallie, "Kinship." See also Binnema, *Common and Contested Ground*, 11; Liette, "Memoir," 364; Perrot, "Memoir," 83–85, 86, 140; Witgen, *An Infinity of Nations*, 122; Havard, *Empire et métissages*, 157–58.
8 Liette, "Memoir," 377, 381; Thwaites, *Jesuit Relations*, 67:171; Raudot, "Memoir," 347.
9 Liette, "Memoir," 384; Rushforth, *Bonds of Alliance*, 35–51; Rushforth, "'A Little Flesh,'" 781; Thwaites, *Jesuit Relations*, 67:173; Raudot, "Memoir," 361.

10 Rushforth, *Bonds of Alliance*, 51–58.
11 Rushforth, "A Little Flesh," 783; Rushforth, *Bonds of Alliance*, 36–37; Raudot, "Memoir," 360.
12 See Sayre, *Les Sauvages Américains*, 283–96; Richter, *The Ordeal of the Longhouse*, 30–74; Rushforth, "A Little Flesh," 783–85; Rushforth, *Bonds of Alliance*, 47.
13 Raudot, "Memoir," 360; DuVal, "Indian Intermarriage."
14 Hennepin, *A New Discovery*, 2:91; Richter, *The Ordeal of the Longhouse*, 69.
15 Raudot, "Memoir," 356; Thwaites, *Jesuit Relations*, 67:169–73; Liette, "Memoir," 383–86; Rushforth, "A Little Flesh," 782.
16 Charlevoix, *Journal of a Voyage*, 187; Thomas Jefferson Farnham, in Thwaites, *Early Western Travels*, 29:392–93.
17 Liette, "Memoir," 377.
18 Liette, "Memoir," 377, 383–84; Raudot, "Memoir," 360, 356.
19 Swedlund, "Contagion, Conflict, and Captivity"; Cameron, "The Effects of Warfare and Captive-Taking"; Kelton, *Epidemics and Enslavement*, 105–7; Thwaites, *Jesuit Relations*, 62:71–73. See also Richter, *The Ordeal of the Longhouse*, 74.
20 Richter, *The Ordeal of the Longhouse*, 66.
21 Perrot, "Memoir," 152.
22 Perrot, "Memoir," 181.
23 Perrot, "Memoir," 191; White, *The Middle Ground*, 3; Thwaites, *Jesuit Relations*, 67:173.
24 J. B. Griffin, "A Hypothesis"; Rushforth, *Bonds of Alliance*, 27, especially 27n14; Thwaites, *Jesuit Relations*, 54:223; Edmunds and Peyser, *The Fox Wars*, 9. For more on the Meskwaki migration, see Buffalo, "Oral History," 3.
25 Emerson and Brown, "The Late Prehistory"; Hämäläinen, *Lakota America*, 15–22.
26 Thwaites, *Jesuit Relations*, 51:29; Edmunds, *The Potawatomis*; Tanner and Hast, *Atlas*. A good summary of the precontact migrations in the region can also be found in Brown and Sasso, "Prelude to History," 212–15.
27 Thwaites, *Jesuit Relations*, 54:223.
28 Thwaites, *Jesuit Relations*, 44:245–47, 18:233–35; Schlesier, "Rethinking the Midewiwin," 2; La Potherie, "History," 293; Witgen, *An Infinity of Nations*, 62; Thwaites, *Jesuit Relations*, 54:219; Perrot, "Memoir," 223.
29 Thwaites, *Jesuit Relations*, 54:231, 51:6.
30 Perrot, "Memoir," 41. For a good summary of the Indigenous population and recent migration history of the western Great Lakes at the time of contact, see Rushforth, *Bonds of Alliance*, 26–29; and Thwaites, *Jesuit Relations*, 60:201. See also White, *The Middle Ground*, 15–16.
31 Perrot, "Memoir," 84–85, 86, 140.
32 Rushforth, *Bonds of Alliance*, 25.
33 La Potherie, "History," 293.
34 Schlesier, "Rethinking the Midewiwin"; Betts, "Pots and Pox"; La Potherie, "History," 293–300.
35 La Potherie, "History," 293, 295–96.
36 Perrot, "Memoir," 159–60, 162–64, 187–88.

37 Perrot, "Memoir," 189–90; Hämäläinen, *Lakota America*, 14–28; Thwaites, *Jesuit Relations*, 54:189–91, 54:221, 55:167.
38 Perrot, "Memoir," 154–55, 181; White, *The Middle Ground*, 9. See also Thwaites, *Jesuit Relations*, 51:39, 51:49–52, 54:215, 55:219; 56:139–47, 60:149–51; WHC 16:65.
39 Thwaites, *Jesuit Relations*, 55:201–3.
40 Thwaites, *Jesuit Relations*, 60:199–201; Witgen, *An Infinity of Nations*, 98; Thwaites, *Jesuit Relations*, 54:223.
41 Thwaites, *Jesuit Relations*, 51:49, 54:191, 55:167.
42 La Potherie, "History," 295.
43 Rushforth, *Bonds of Alliance*, 29.
44 Blakeslee, "The Origin and Spread"; Gundersen, "'Catlinite,'" 560–62.
45 Perrot, "Memoir," 185–86; Thwaites, *Jesuit Relations*, 51:47–49; Lahontan, *New Voyages*, 1:75–76; Perrot, "Memoir," 185; Thwaites, *Jesuit Relations*, 64:123.
46 Blakeslee, "The Origin and Spread," 759; Hall, *An Archaeology of the Soul*, chaps. 1, 6–7; Carayon, *Eloquence Embodied*, 406–15; Rushforth, *Bonds of Alliance*, 25–34. For further discussion of the Calumet and its creation of fictive kinship, see R. White, *The Middle Ground*, 16, 20–22. See also Thwaites, *Jesuit Relations*, 64:123; Perrot, "History," 185; and Thwaites, *Jesuit Relations*, 58:97.
47 Thwaites, *Jesuit Relations*, 59:121, 59:129–35.
48 Rushforth, *Bonds of Alliance*, 33. For other examples, see, for instance, Dumont du Montigny, *Mémoires*, 193–95; See also the "La Harpe Narrative" in Margry, *Découvertes*, 6:292.
49 Rushforth, "A Little Flesh We Offer You," 786.
50 Thwaites, *Jesuit Relations*, 59:135, 59:121; Rushforth, *Bonds of Alliance*, 34; Rushforth, "A Little Flesh," 785.
51 Rushforth, *Bonds of Alliance*, 29, 47.
52 Raudot, "Memoir," 360.
53 La Potherie, "History," 187. For the Sioux strategy of releasing captives, see Raudot, "Memoir," 378; Perrot, in WHC 16:30–31; Rushforth, "A Little Flesh," 786; Perrot, "History," 187–88; and Thwaites, *Jesuit Relations*, 54:237, 55:169.
54 Dollier de Casson, in Kellogg, *Early Narratives*, 181–82; Rushforth, "Slavery," 56; Rushforth, *Bonds of Alliance*, 30.
55 Thwaites, *Jesuit Relations*, 54:225.
56 Thwaites, *Jesuit Relations*, 54:177, 59:135; Margry, *Découvertes*, 6:21; Perrot, "Memoir," 185–86, 90.
57 For the usefulness of these analytics, as opposed to "confederacy" or "alliance," see Binnema, *Common and Contested Ground*, 15.
58 Rushforth, *Bonds of Alliance*, 23.
59 R. White, *The Middle Ground*, 14.
60 Thwaites, *Jesuit Relations*, 59:125–27.
61 Thwaites, *Jesuit Relations*, 54:189, 60:161. See the Shawnee villagers quoted in R. White, *The Middle Ground*, 49. See also R. La Salle, *Relation of the Discoveries*, 213.
62 Edmunds and Peyser, *The Fox Wars*, 23.

63 Thwaites, *Jesuit Relations*, 54:225.
64 Daniel de Rémy, sieur de Courcelle, quoted in Edmunds and Peyser, *The Fox Wars*, 23.
65 Hennepin, *A New Discovery*, 1:135.

5. THE GREAT BISON ACCELERATION

1 Thwaites, *Jesuit Relations*, 67:165–67.
2 Thwaites, *Jesuit Relations*, 65:75, 65:229–31.
3 Thwaites, *Jesuit Relations*, 67:169–73.
4 For "glory," see Thwaites, *Jesuit Relations*, 67:169–71.
5 For similar impressions, see Crouch, *Nobility Lost*. For "honor" in Indigenous warfare as the French broadly understood it, see Sayre, *Les Sauvages Américains*, 268–83.
6 See, for instance, Smits, "The 'Squaw Drudge.'"
7 Sleeper-Smith, *Indian Women and French Men*, 23; Sleeper-Smith, "Women, Kin, and Catholicism," 426–29.
8 Thwaites, *Jesuit Relations*, 67:167; Liette, "Memoir," 318.
9 For "held together," see Law, "Technology and Heterogeneous Engineering."
10 La Salle said there were 460 cabins, four to five fires per cabin, and two families per fire. This was a huge population. See also Thwaites, *Jesuit Relations*, 67:161.
11 For "Grown somewhat scarce," see La Salle, quoted in Hennepin, *A New Discovery* 2:627.
12 Thwaites, *Jesuit Relations*, 67:167–69.
13 Hennepin, *A New Discovery*, 2:563.
14 De Tonti, *Relation of Henri De Tonty*, 29; Margry, *Découvertes*, 1:582.
15 St. Cosme, "Voyage of St. Cosme," 350.
16 Thwaites, *Jesuit Relations*, 55:195.
17 Liette, "Memoir," 308; Hennepin, *A New Discovery*, 1:154.
18 Perrot, "Memoir," 120, 125.
19 Raudot, "Memoir," 409.
20 Hennepin, *A New Discovery*, 2:563; Gross and Howard, "Colbert, La Salle," 90n15; La Salle to his men, 1683, ISHL 23:40.
21 Isenberg, *The Destruction of the Bison*, 95, 99–100.
22 Thwaites, *Jesuit Relations*, 59:111; Hennepin, *A New Discovery*, 2:645.
23 Hennepin, *A New Discovery*, 2:520–21; Thwaites, *Jesuit Relations*, 65:73–75.
24 Hennepin, *A New Discovery*, 1:149; Liette, "Memoir," 318; Perrot, "Memoir," 124; Thwaites, *Jesuit Relations*, 65:73–75.
25 Thwaites, *Jesuit Relations*, 65:75, 66: 251; Perrot, "Memoir," 122.
26 Hennepin, *A New Discovery*, 1:242; Perrot, "Memoir," 124. For the "entrails," see Hennepin, *A New Discovery*, 2:520–21. See also Krech, *The Ecological Indian*, 133–35.
27 Hennepin, *A New Discovery*, 2:520–21; Thwaites, *Jesuit Relations*, 65:73–75; Raudot, "Memoir," 364; Liette, "Memoir," 318. See also Hennepin, *A New Discovery*, 2:490.

28 Raudot, "Memoir," 407–8. See also Anonymous, "Relation de La Louisianne," 202–3; and Liette, "Memoir," 312. Sometimes "old men" joined in to perform some of the butchering.

29 Hennepin, *A Description of Louisiana*, 144; Hennepin, *A New Discovery*, 2:490.

30 Anonymous, "Relation de La Louisianne," 147; Thwaites, *Jesuit Relations*, 67:165, 66:227–31; Perrot, "Memoir," 75. See also Thwaites, *Jesuit Relations*, 68:201–3.

31 For the normal division among prairie groups generally, see Perrot, "Memoir," 74–76. For a comparable situation, see T. Edwards, *Osage Women and Empire*, chap. 1.

32 Thwaites, *Jesuit Relations*, 65:73, 66:231.

33 Thwaites, *Jesuit Relations*, 67:165–67.

34 Perrot, "Memoir," 124; Hennepin, *A New Discovery*, 1:147; Margry, *Découvertes*, 5:87.

35 Raudot, "Memoir," 408; Perrot, "Memoir," 120–21; Liette, "Memoir," 328, 376–77.

36 Perrot, "Memoir," 135; and Cadillac's description of the Miamis in Margry, *Découvertes*, 5:86.

37 R. La Salle, *Relation of the Discoveries*, 145.

38 Liette, "Memoir," 355; Isenberg, *The Destruction of the Bison*, 100–101. See also Hämäläinen, *The Comanche Empire*, 248; R. White, *The Roots of Dependency*, 40–42; and Liette, "Memoir," 331.

39 For multiple wives, see Hennepin, *A New Discovery*, 1:167; Liette, "Memoir," 355; R. La Salle, *Relation of the Discoveries*, 295; Margry, *Découvertes*, 2:98–99. See also Margry, *Découvertes*, 1:542; and Perrot, "Memoir," 72–73. Importantly, when it came to this extreme of divorce, the departing wife would also be entitled to his property.

40 Rushforth, *Bonds of Alliance*, 59; Perrot, "Memoir," 64–65, 69, 75.

41 Gravier, *Kaskakia-to-French Dictionary*, 509, 346, 335, 394, 37, 51, ILDA; Rushforth, *Bonds of Alliance*, 57–59.

42 Unidentified Frenchman quoted in Hauser, "The Berdache," 55; D'Artaguiette, "Journal," 73; Liette, "Memoir," 333, 353; Raudot, "Memoir," 394, Pénicaut, *Fleur de Lys and Calumet*, 140; Thwaites, *Jesuit Relations*, 65:66; Hennepin, *A New Discovery*, 2:480–81. For similar devaluing of "wives," which were likely slaves, see Perrot, "Memoir," 144; and Cadillac, in WHC 16:361–62.

43 For instance, Liette, "Memoir," 335–37; Hennepin, *New Discovery*, 2:652; and R. La Salle, *Relation of the Discoveries*, 145. See also description of scalping of Illinois "wives" in Lallement, "A Copy of a Letter." See also Rushforth, *Bonds of Alliance*, 69.

44 Raudot, "Memoir," 347–48, 351, 403–5; Liette, "Memoir," 377, 380. For the complex word *mataché* see Robert Vezina's etymology in Trudeau, *A Fur Trader*, 541–42; and Thwaites, *Jesuit Relations*, 67:171.

45 Liette, "Memoir," 377; Hennepin, *New Discovery*, 1:175, 2:664. See also R. La Salle, *Relation of the Discoveries*, 137; Margry, *Découvertes*, 2:324–26; Ekberg, *Stealing Indian Women*, 12; de Tonti, "Memoir on LaSalle's Discoveries," 303, 313;

Hauser, "Warfare and the Illinois," 379; Barr, "From Captives to Slaves," 23–24; and Raudot, "Memoir," 403–4.

46 Raudot, "Memoir," 404.

47 Liette, "Memoir," 387, 381; La Potherie, "History," 300. See also Margry, *Découvertes*, 5:492–93.

48 R. La Salle, *Relation of the Discoveries*, 145; Margry *Découvertes*, 2: 98–99. See also Thwaites, *Jesuit Relations*, 67:171–73.

49 Lahontan, *New Voyages*, 1:94, 1:106, 2:541, 2:575. Liette confirms that slaves were often part of a bride price in Illinois. Liette, "Memoir," 331.

50 Rushforth, *Bonds of Alliance*, 55; La Salle, ISHL 23:10. Although the practice of captivity was changing significantly in the contact period for many groups, the significance of labor in the Illinois' system at this time made it distinctive. See, for instance, Pesantubbee, *Choctaw Women in a Chaotic World*, chap. 2.

51 Jean Lamberville, in Thwaites, *Jesuit Relations*, 60:185.

52 Rushforth, *Bonds of Alliance*, 36–37, 57–59, 389.

53 Margry, *Découvertes*, 5: 94–95; Rushforth, *Bonds of Alliance*, 58–59; Thwaites, *Jesuit Relations*, 64: 219; Liette, "Memoir," 318.

54 Hennepin, *A New Discovery*, 2:627; La Salle on Illinois Country, 1680, ISHL 23:4. For the elimination of "buffer zones" as a refuge for game populations, see Lahontan, *New Voyages*, 1:193; and R. La Salle, *Relation of the Discoveries*, 271.

55 Stambaugh et al., "Drought Duration"; Morrissey, "Climate, Ecology, and History."

56 La Source, "Voyage Up the Mississippi," 83.

57 St. Cosme, "Voyage of St. Cosme," 350.

58 Lott, *American Bison*; Hämäläinen, "The Politics of Grass."

59 Hennepin, *A New Discovery*, 1:148.

60 Gonella, "Myaamia Ethnobotany," 29.

61 Liette, "Memoir," 326.

62 McCafferty, "The Illinois Place Name 'Pimitéoui,'" 177.

63 Thwaites, *Jesuit Relations*, 64:171.

6. HIDING IN THE TALLGRASS

1 Thwaites, *Jesuit Relations*, 66:269–73, 66:281, 66:287.

2 Havard, *The Great Peace*.

3 Fortier and Chaput, "A Historical Reexamination"; Caldwell, "Charles Juchereau de St. Denys."

4 Fauteux, *Essai sur l'industrie*, 2:410.

5 Ironstrack, "The Good Path: Part 1"; Sutterfield, "Aciipihkahki"; Margry, *Découvertes*, 2:244; Pénicaut, *Fleur de Lys*, 142; Thwaites, *Jesuit Relations*, 65:111.

6 For sparse population, see Thwaites, *Jesuit Relations*, 66:229. For the Wabash, see Diderot and d'Alembert, *Encyclopédie*, 11:697.

7 Bergier quoted in Palm, *The Jesuit Missions*, 31; Thwaites, *Jesuit Relations*, 65:99–103.

8 Margry, *Découvertes*, 4:595–97.

9 Gross and Howard, "Colbert, La Salle"; Margry, *Découvertes*, 3:22, 2:78–80; R. La Salle, *Relation of the Discoveries*, 15.
10 Margry, *Découvertes*, 2:81, 2:244, 3:22, 2:260; Gross and Howard, "Colbert, La Salle."
11 Margry, *Découvertes*, 2:260, 2:244.
12 Margry, *Découvertes*, 4:370–77.
13 Fortier and Chaput, "A Historical Reexamination," 385–86; Margry, *Découvertes*, 5:356–60.
14 Margry, *Découvertes*, 5:351–53; Fortier and Chaput, "A Historical Reexamination," 389–90.
15 Grobbel, "Tanning and Tanneries," 121–22; Welsh, "Tanning in the United States," 3, 12–13.
16 Welsh, "Tanning in the United States," chap. 2, esp. 21–22; Grobbel, "Tanning and Tanneries," 122–25; Comité de salut public, *Programmes des cours*, 122.
17 Riello, "Nature, Production and Regulation," 75–76.
18 Caldwell, "Charles Juchereau de St. Denys," 568n24; Fauteux, *Essai sur l'industrie*, 2:406–10.
19 Writing of the buffalo as both a type of animal but also a class of tanned leather, one economist of the early nineteenth century said: "The skin, strong, smooth and elastic, receives different preparations; it is above all employed under the name of 'buffle' after having been chamoised." See Lasteyrie, *Histoire naturelle*, 42.
20 Durival, "Habillement, Équipement," in Diderot, *Encyclopédie*, 8:7; Louis XV, *Ordonnance du Roi, concernant la cavalerie*, 39; Fauteux, *Essai sur l'industrie*, 2:405–6; Comité de salut public, *Programmes des cours*.
21 Michaud, "Manufacture de buffles," 6.
22 See, for instance, "du travail du buffles" in Roland de la Platière, *Encyclopedie Méthodique*, chap. 3.
23 Le Boullenger, *Myaamia-Illinois Dictionary*, 320.
24 Weyhing, "Le Sueur in the Sioux Country"; Pénicaut, *Fleur de Lys*, 32–49.
25 Pénicaut, *Fleur de Lys*, 62–63; Caldwell, "Charles Juchereau de St. Denys," 575; Fortier and Chaput, "A Historical Reexamination," 392–93.
26 Thwaites, *Jesuit Relations*, 66:39–41.
27 Fortier and Chaput, "A Historical Reexamination," 398.
28 Bourgmont reportedly joined the Mascoutens, who settled in a village for eighteen months in order to supply the tannery. Veniard, "An Exact Description of Louisiana," 33.
29 Charlevoix, *Journal*, 1:188.
30 Baillargeon, *North American Aboriginal Hide Tanning*, 15.
31 Gravier and Largiller, *Kaskaskia-to-French Dictionary*, 254, 75, 90, ILDA.
32 Pinet, *Miami-Illinois Dictionary*, 77, ILDA.
33 Gravier and Largillier, *Kaskaskia-to-French Dictionary*, 465, ILDA.
34 Baillargeon, *North American Aboriginal Hide Tanning*, 13–24.
35 See "The Tanner as Artist" in Baillargeon, *North American Aboriginal Hide Tanning*.

36 Baillargeon, *North American Aboriginal Hide Tanning*, 4; Harrod, *The Animals Came Dancing*; J. E. Brown, *Animals of the Soul*.
37 Baillargeon, *North American Aboriginal Hide Tanning*, 5–6.
38 Habicht-Mauche, "The Shifting Role," 42.
39 Jablow, *The Cheyenne*, 20.
40 Frink and Weedman eds., *Gender and Hide Production*; Habicht-Mauche, "Women on the Edge"; Isenberg, *The Destruction of the Bison*, 93–100.
41 Habicht-Mauche, "The Shifting Role," 37–38, 42.
42 See Iberville's journal in Margry, *Découvertes*, 4:406.
43 Margry, *Découvertes*, 4:544, 561.
44 St. Cosme, 1698, quoted in L. Jones, *The Shattered Cross*, 72–73; Shea, *Relation de la mission*, 61. See also Iberville's "Mémoire of 1701" in Margry, *Découvertes*, 4:544.
45 Margry, *Découvertes*, 4:520.
46 Margry, *Découvertes*, 4:406, 457, 516.
47 Margry, *Découvertes*, 4:594.
48 Margry, *Découvertes*, 4:517.
49 St. Cosme, "St. Cosme to the Bishop of Quebec," 85.
50 Raudot, "Memoir," 402–5; For evidence that these raids were ongoing through the early 1700s, see Pénicaut, *Fleur de Lys*, 122–23.
51 Kelton, *Epidemics and Enslavement*, chap. 3.
52 Kelton, *Epidemics and Enslavement*, 102, 149.
53 DuVal, *The Native Ground*, 78; Shea, *Relation de la mission*, 72–73; La Source, "Voyage Up the Mississippi," 79, 81–82.
54 Kelton, *Epidemics and Enslavement*, 150.
55 Thwaites, *Jesuit Relations*, 66:237–41; Blasingham, "The Illinois Indians," 383.
56 Thwaites, *Jesuit Relations*, 66:247, 251, 257.
57 For the older way of thinking about Indian populations as "doomed to die" because of the biological inevitability of disease and lack of immunity, see White, *The Middle Ground*, 41. For "immunological determinism" see D. Jones, "Virgin Soils Revisited."
58 Milner, "Population Decline"; Betts, "Pots and Pox."
59 Edwards and Kelton, "Germs, Genocides," 74.
60 This metaphor is from DeCoster, "Tangled in the Web."
61 Extract of a letter from Bienville to Minister, September 1704, in Margry, *Découvertes*, 5:368–69.
62 Pénicaut, *Fleur de Lys*, 76–78.
63 At least eight to nine thousand bison hides departed Illinois Country, and possibly as many as fifteen thousand. Fortier and Chaput, "A Historical Reexamination," 398; Fauteux, *Essai l'industrie*, 1:417.
64 Thwaites, *Jesuit Relations*, 66:291.
65 For this older theme of "dependency," see Hauser, "The Illinois Indian Tribe"; and R. White, *The Roots of Dependency*.

7. WAR

1 Rushforth, "Slavery"; Edmunds and Peyser, *The Fox Wars*, 41.
2 For a good reflection on these early works, see R. White, *The Middle Ground*, 154n19.
3 White, *Middle Ground*, chap. 4; Edmunds and Peyser, *The Fox Wars*. More recently, Brett Rushforth told a truly Native-centered account in "Slavery."
4 Weyhing, "'Gascon Exagerrations'"; R. White, *The Middle Ground*, 150–52; For strategies behind Detroit, see "Mémoire adresse au Comte de Maurepas" in Margry, *Découvertes*, 5:138–53.
5 Edmunds and Peyser, *The Fox Wars*, 55–56.
6 De Léry (1712), in WHC 16:293.
7 R. White, *The Middle Ground*, 151.
8 Dubuisson (1712), in WHC 16:272.
9 Dubuisson (1712), in WHC 16:278.
10 Marest to Vaudreuil, in WHC 16:289.
11 Edmunds and Peyser, *The Fox Wars*, 75. According to Marest, Fox losses exceeded eight hundred captured. See Marest to Vaudreuil, in WHC 16:288–89.
12 R. White, *The Middle Ground*, 161–64.
13 Rushforth, "Slavery," 66.
14 R. White, *The Middle Ground*, 160; Vaudreuil (1714), in WHC 16:306; Vaudreuil and Bégon to the Minister, in WHC 16:298; de Ramezay to the Minister, in WHC 16:300.
15 Vaudreuil and Bégon (1714), in WHC 16:304.
16 Thwaites, *Jesuit Relations*, 66:291; Ramezay and Bégon, in WHC 16:317, 319; Rushforth, "Slavery," 66–68.
17 Edmunds and Peyser, *The Fox Wars*, 80–81; Ramezay (1715), in WHC 16:323; Blasingham, "The Illinois Indians," 383–84; Palm, *Jesuit Missions*, 45; Vaudreuil to Marine (1716), in WHC 16:341–42.
18 Edmunds and Peyser, *The Fox Wars*, 82; Ramezay to the Minister, in WHC 16: 300.
19 Extract from letter of Louvigny to Louis Alexandre de Bourbon, in WHC 16:348–49.
20 Proceedings in French Council of Marine (1716), in WHC 16:338–40.
21 Rushforth, "Slavery," 67–68; Vaudreuil to Marine (1716), in WHC 16:343.
22 La Mothe Cadillac (1718), in WHC 16:361; Edmunds and Peyser, *The Fox Wars*, 94; Le Boullenger and Kereben to Vaudreuil, in WHC 16:455; extract from letter of Louvigny to Louis Alexandre de Bourbon (1717), in WHC 16:348–49; Sabrevois, in WHC 16:371–72; Vaudreuil (1720), in WHC 16:392.
23 Edmunds and Peyser, *The Fox Wars*, 92–93; Anakapita, Massauga, and Jonachin recorded by DuTisné, in WHC 16:459–62.
24 Dumont de Montigny, *The Memoir of Lieutenant Dumont*, 372–404.
25 Veniard, "Exact Description of Louisiana."
26 Veniard, "Journal of the Voyage," 131.
27 Ellis and Steen, "An Indian Delegation," 395.

28 Bourgmont, "Journal of the Voyage," January 11, February 18; Blaine, *The Ioway Indians*, 31. See also Boisbriant (1721), Archives de la Guerre, 2592, fols. 97–99.
29 Vaudreuil to Council (1718), in WHC 16:381.
30 R. White, *The Middle Ground*, 161–64; Charlevoix, *Journal*, 2:198–202; Edmunds and Peyser, *The Fox Wars*, 97; D'Artaguiette, "Journal," 83; Charlevoix, *History and General Description of New France*, 5:130.
31 D'Artaguiette, "Journal," 70–71, 83–84; Thwaites, *Jesuit Relations*, 67:283.
32 D'Artaguiette, "Journal," 71–73.
33 Charlevoix, *Journal*, 2:190; R. White, *The Middle Ground*, 166; W. Jones, *Fox Texts*, 9–11; Lallement, "A Copy of a Letter"; De Lignery (1724), in WHC 16:445; Edmunds and Peyser, *The Fox Wars*, 100.
34 Villedonné and Mesagier (1724), in WHC 16:447.
35 Anakapita, Massauga, and Jonachin recorded by DuTisné, in WHC 16:457–63.
36 Missionaries at Kaskaskia (1725), in WHC 16:454; Du Tisné to Vaudreuil, January 14, 1725, in WHC 16:450–51.
37 Vaudreuil to Boisbriant (May 20, 1724) in WHC 16:441–42; De Lignery to Boisbriant, in WHC 16:445.
38 See the children Michel and Louis Loisel, in Faribault-Beauregard, *La Population des forts Français*, 237.
39 Anakapita, Massauga, and Jonachin recorded by DuTisné, in WHC 16:458; Le Boullenger and Kereben to Vaudreuil (January 10, 1725) in WHC 16:455; Du Tisné (1725), in WHC 16:451.
40 Du Tisné (1725), in WHC 16:452; Anakapita, Massauga, and Jonachin recorded by DuTisné, in WHC 16:456, 462.
41 Anakapita, Massauga, and Jonachin recorded by DuTisné, in WHC 16:462.
42 Edmunds and Peyser, *The Fox Wars*, 102; Missionaries at Kaskaskia, in WHC 16:453.
43 Edmunds and Peyser, *The Fox Wars*, 107; Vaudreuil to Boisbriant, in WHC 16:443.
44 See, for instance, Peyser, "The 1730 Siege," 152; and WHC 17:111–12, 17:115.
45 Peyser, "The Fate of the Fox Survivors," 87; Edmunds and Peyser, *The Fox Wars*, 104–7. See also WHC 17:5; and De Lignery's Peace, in WHC 16:466.
46 Edmunds and Peyser, *The Fox Wars*, 111; WHC 3:161–62.
47 R. White, *The Middle Ground*, 166. Another estimate was that there were 1,200 Indigenous soldiers and 450 French soldiers. De Lignery's Report (1728), in WHC 17:32–33.
48 Boucherville's narrative, in WHC 17:51; Edmunds and Peyser, *The Fox Wars*, 129; Peyser, "The Fate of the Fox Survivors," 90–91.
49 Peyser, "The Fate of the Fox Survivors," 91; Edmunds and Peyser, *The Fox Wars*, 137, 144–45; WHC 17:110–15.
50 Edmunds and Peyser, *The Fox Wars*, 139.
51 For the best accounts, see Edmunds and Peyser, *The Fox Wars*; Peyser, "The 1730 Siege"; Stelle and Hargrave, "Messages in a Map"; and Stelle, "History and Archaeology."
52 WHC 17:111, 17:115.

53 WHC 17:111.
54 Périer to Maurepas, March 25, 1731, in *Mississippi Provincial Archives*, 4:72.
55 WHC 16:459–62.
56 A maritime accident caused these intended gifts to be lost. Ellis and Steen, "An Indian Delegation," 395.
57 WHC 17:148, 17:152–53; R. White, *The Middle Ground*, 168–71.
58 WHC 17:148–54; Peyser, "The Fate of the Fox Survivors," 99–100.
59 R. White, *The Middle Ground*, 169.
60 Quoted in R. White, *The Middle Ground*, 173.
61 Blasingham, "The Illinois Indians," 378; R. White, *The Middle Ground*, 169n38.

CONCLUSION

1 Havard, "Un Américain à Rochefort," 217–19.
2 Gravier, *Kaskaskia-to-French Dictionary*, 199, ILDA.
3 Krutak, *Tattoo Traditions*, 202–3; Brasser, "Notes on a Recently Discovered Indian Shirt," 46–55.
4 Havard, "Un Américain à Rochefort," 149.
5 De Tonti, *Relation of Henri de Tonty*, 29; Lahontan, *New Voyages*, 2:415; Cadillac, in WHC 16:361; Charlevoix, *Journal*, 2:193.
6 Parmenter, *The Edge of the Woods*, xxxiv; Hämäläinen and Truett, "On Borderlands," 346.
7 Shriver, "Thoughts on COVID Neehseehpineenki."
8 Ostler and Shoemaker, "Settler Colonialism."
9 Lallement, "A Copy of a Letter."
10 Ellis and Steen, "An Indian Delegation in France."
11 Anonymous, "Relation de La Louisianne," 201–4.
12 Morris, "How to Prepare Buffalo."
13 Smalley, *Wild by Nature*.
14 Morrissey, "'The Country Is Greatly Injured.'"
15 Anonymous, "Relation de La Louisianne," 38.
16 Bartram, "Journal," 144.
17 Anonymous, "Relation de La Louisianne," 194–95.
18 Stelle and Hargrave, "Messages in a Map"; Stelle, "History and Archaeology."
19 See especially the extraordinary archaeological investigations led by Lenville J. Stelle.

BIBLIOGRAPHY

Allen, Matthew S., and Michael W. Palmer. "Fire History of a Prairie/Forest Boundary: More than 250 Years of Frequent Fire in a North American Tallgrass Prairie." *Journal of Vegetation Science* 22, no. 3 (June 2011): 436–44.

Allred, Brady W., Samuel D. Fuhlendorf, David M. Engle, and R. Dwayne Elmore. "Ungulate Preference for Burned Patches Reveals Strength of Fire-Grazing Interaction." *Ecology and Evolution* 1, no. 2 (2011): 132–44.

Anderson, David G., and Marvin T. Smith. "Pre-Contact: The Evidence from Archaeology." In *A Companion to Colonial America*, edited by Daniel Vickers, 1–24. New York: John Wiley & Sons, 2007.

Anderson, Kat. *Tending the Wild: Native American Knowledge and the Management of California's Natural Resources*. Berkeley: University of California Press, 2005.

Anderson, Roger C. "The Eastern Prairie-Forest Transition—An Overview." In *Proceedings of the Eighth North American Prairie Conference*, 86–92. Kalamazoo: Western Michigan University, 1983.

———. "Evolution and Origin of the Central Grassland of North America: Climate, Fire, and Mammalian Grazers." *Journal of the Torrey Botanical Society* 133, no. 4 (October 1, 2006): 626–47.

———. "The Historic Role of Fire in the North American Grassland." In *Fire in North American Tallgrass Prairies*, edited by Scott L. Collins and Linda L. Wallace, 8–18. Norman: University of Oklahoma Press, 1990.

———. "Presettlement Forest of Illinois." In *Proceedings of the Oak Woods Management Workshop*, edited by J. E. Ebinger, Gerould Wilhelm, and George V. Burger, 9–19. Charleston: Eastern Illinois University, 1991.

Anderson, Roger C., and M. Rebecca Anderson. "The Presettlement Vegetation of Williamson County, Illinois." *Castanea* 40, no. 4 (1975): 345–63.

Andrews, Thomas G. *Coyote Valley: Deep History in the High Rockies*. Cambridge, MA: Harvard University Press, 2015.

———. *Killing for Coal: America's Deadliest Labor War.* Cambridge, MA: Harvard University Press, 2008.

Anonymous. "Relation de La Louisianne." 1735. Ayer 530. Newberry Library.

Archives de la Guerre. Centre Historique des Archives Nationales, Paris, France.

Arthur, George W. *An Introduction to the Ecology of Early Historic Communal Bison Hunting among the Northern Plains Indians.* Ottawa: National Museums of Canada, 1975.

Asch, David L., and Nancy B. Asch. "Woodland Period Archeobotany of the Napoleon Hollow Site." In *Woodland Period Occupations of the Napoleon Hollow Site in the Lower Illinois Valley*, edited by Michael D. Wiant and Charles R. McGimsey, 427–512. Kampsville, IL: Center for American Archeology, 1986.

Axelrod, Daniel I. "Rise of the Grassland Biome, Central North America." *Botanical Review* 51, no. 2 (1985): 163–201.

Baerreis, David A., Reid A. Bryson, and John E. Kutzbach. "Climate and Culture in the Western Great Lakes Region." *Midcontinental Journal of Archaeology* 1, no. 1 (1976): 39–57.

Bailey, Robert. "Description of the Ecoregions of the United States." Accessed April 17, 2019. https://www.fs.fed.us/land/ecosysmgmt/.

Baillargeon, Morgan. *North American Aboriginal Hide Tanning: The Act of Transformation and Revival.* Gatineau, Québec: Canadian Museum of Civilization, 2011.

Baker, Richard G., Louis J. Maher, Craig A. Chumbley, and Kent L. Van Zant. "Patterns of Holocene Environmental Change in the Midwestern United States." *Quaternary Research* 37, no. 3 (May 1, 1992): 379–89.

Baldwin, Daryl, and David J. Costa. *Myaamia Neehi Peewaalia Kaloosioni Mahsinaakani: A Miami-Peoria Dictionary.* Miami, OK: Miami Tribe of Oklahoma, 2005.

Balée, William L., and Clark L. Erickson, eds. *Time and Complexity in Historical Ecology: Studies in the Neotropical Lowlands.* New York: Columbia University Press, 2006.

Bamforth, Douglas B. *Ecology and Human Organization on the Great Plains.* New York: Plenum, 1988

Barbour, Michael. "Ecological Fragmentation in the Fifties." In *Uncommon Ground: Rethinking the Human Place in Nature*, edited by William Cronon, 233–55. New York: Norton, 1996.

Barr, Juliana. "From Captives to Slaves: Commodifying Women in the Borderlands." *Journal of American History* 92, no. 1 (2005): 19–46.

———. "There's No Such Thing as 'Prehistory': What the Longue Durée of Caddo and Pueblo History Tells Us about Colonial America." *William and Mary Quarterly* 74, no. 2 (2017): 203–40.

Barsh, Russel Lawrence, and Chantelle Marlor. "Driving Bison and Blackfoot Science." *Human Ecology* 31, no. 4 (2003): 571–93.

Bartram, William. "Journal—Observations on the Creek and Cherokee Indians." In *William Bartram on the Southeastern Indians*, edited by Kathryn E. Holland Braund and Gregory A. Waselkov, 133–86. Lincoln: University of Nebraska Press, 1995.

Benn, David W. "Hawks, Serpents, and Bird-Men: Emergence of the Oneota Mode of Production." *Plains Anthropologist* 34, no. 125 (1989): 233–60.

Bennett, Jane. *Vibrant Matter: A Political Ecology of Things*. Durham, NC: Duke University Press, 2010.

Benson, Larry V., Michael S. Berry, Edward A. Jolie, Jerry D. Spangler, David W. Stahle, and Eugene M. Hattori. "Possible Impacts of Early-11th-, Middle-12th-, and Late-13th-Century Droughts on Western Native Americans and the Mississippian Cahokians." *Quaternary Science Reviews* 26, no. 3–4 (2007): 336–50.

Benson, Larry V., Timothy R. Pauketat, and Edward R. Cook. "Cahokia's Boom and Bust in the Context of Climate Change." *American Antiquity* 74, no. 3 (2009): 467–83.

Betts, Colin M. "Pots and Pox: The Identification of Protohistoric Epidemics in the Upper Mississippi Valley." *American Antiquity* 71, no. 2 (2006): 233–59.

Betz, Robert F. *The Prairie of the Illinois Country*. Westmont, IL: DPM Ink, 2011.

Binford, Lewis R. "Willow Smoke and Dogs' Tails: Hunter-Gatherer Settlement Systems and Archaeological Site Formation." *American Antiquity* 45, no. 1 (1980): 4–20.

Binnema, Theodore. *Common and Contested Ground: A Human and Environmental History of the Northwestern Plains*. Norman: University of Oklahoma Press, 2001.

Blaine, Martha Royce. *The Ioway Indians*. University of Oklahoma Press, 1995.

Blakeslee, Donald J. "The Origin and Spread of the Calumet Ceremony." *American Antiquity* 46, no. 4 (1981): 759–68.

Blasingham, Emily Jane. "The Illinois Indians, 1634–1800: A Study of Depopulation." *Ethnohistory* 3, no 1–2 (Summer/Fall 1956): 193–224, 361–412.

Bluhm, Elaine A., and Allen Liss. "The Anker Site." *Chicago Area Archaeology, Illinois Archaeological Survey Bulletin* 3 (1961): 89–138.

Boehm, Andrew Ray. "Were Bison Predictable Prey?: Using Stable Isotopes to Examine Early Holocene Bison Mobility on the Central Great Plains." PhD diss., Southern Methodist University, 2016.

Boggess, William R., and James W. Geis. "The Prairie Peninsula: Its Origin and Significance in the Vegetational History of Central Illinois." In *The Quaternary of Illinois: A Symposium in Observance of the Centennial of the University of Illinois*, edited by Robert E. Bergstrom, 89–95. Urbana: University of Illinois College of Agriculture, 1968.

Borchert, John R. "The Climate of the Central North American Grassland." *Annals of the Association of American Geographers* 40, no. 1 (1950): 1–39.

Bossu, Jean-Benard. *Travels through That Part of North America Formerly Called Louisiana*. Translated by Johann Reinhold Forster. London: T. Davies, 1771.

Boszhardt, Robert F. "Turquoise, Rasps and Heartlines: The Oneota Bison Pull." In *Mounds, Modoc, and Mesoamerica: Papers in Honor of Melvin L. Fowler*, edited by Steven R. Ahler, 361–73. Springfield: Illinois State Museum, 2000.

Boszhardt, Robert F., and Joelle McCarthy. "Oneota End Scrapers and Experiments in Hide Dressing: An Analysis from the La Crosse Locality." *Midcontinental Journal of Archaeology* 24, no. 2 (October 1999): 177–77.

Bourgmont, Étienne Veniard de. "An Exact Description of Louisiana, 1714." Edited by Marcel Giraud. Translated by Max W. Myers. *Bulletin of the Missouri Historical Society* 15 (1958): 3–19.

———. "Exact Description of Louisiana." In *Bourgmont, Explorer of the Missouri, 1698–1725*, edited and translated by Frank Norall, 99–112. Lincoln: University of Nebraska Press, 1988.

———. "Journal of the Voyage to the Padoucas." In *Bourgmont, Explorer of the Missouri, 1698–1725*, edited and translated by Frank Norall, 125–61. Lincoln: University of Nebraska Press, 1988.

———. "The Route to Be Taken to Ascend the Missouri River." In *Bourgmont, Explorer of the Missouri, 1698–1725*, edited and translated by Frank Norall, 113–23. Lincoln: University of Nebraska Press, 1988.

Bramble, Dennis M., and Daniel E. Lieberman. "Endurance Running and the Evolution of Homo." *Nature* 432, no. 7015 (November 2004): 345–52.

Breen, Benjamin. "'The Elks Are Our Horses': Animals and Domestication in the New France Borderlands." *Journal of Early American History* 3, no. 2–3 (2013): 181–206.

Brasser, Ted J. "Notes on a Recently Discovered Indian Shirt from New France." *American Indian Art* 24 (Spring 1999): 46–55.

Braudel, Fernand. *The Mediterranean and the Mediterranean World in the Age of Philip II*. 2 vols. Berkeley: University of California Press, 1996.

Brooke, John L. "Ecology." In *A Companion to Colonial America*, edited by Daniel Vickers, 44–75. New York: John Wiley & Sons, 2007.

Brooks, James F. *Captives and Cousins: Slavery, Kinship, and Community in the Southwest Borderlands*. Chapel Hill: University of North Carolina Press, 2001.

Brooks, Lisa. "Awikhigawôgan Ta Pildowi Ôjmowôgan: Mapping a New History." *William and Mary Quarterly* 75, no. 2 (2018): 259–94.

Brown, James A. *Aboriginal Cultural Adaptations in the Midwestern Prairies*. New York: Garland, 1991.

———. "The Prairie Peninsula: An Interaction Area in the Eastern United States." PhD diss., University of Chicago, 1965.

———. "What Kind of Economy Did the Oneota Have?." In *Oneota Studies*, edited by Guy E. Gibbon, 107–12. Minneapolis: University of Minnesota Press, 1982.

Brown, James A., and Robert F. Sasso. "Prelude to History on the Eastern Prairies." In *Societies in Eclipse: Archaeology of the Eastern Woodlands Indians, AD 1400–1700*, edited by David S. Brose, C. Wesley Cowan, and Robert C. Mainfort, 205–28. Washington, DC: Smithsonian Institution Press, 2001.

Brown, James H., James F. Gillooly, Andrew P. Allen, Van M. Savage, and Geoffrey B. West. "Toward a Metabolic Theory of Ecology." *Ecology* 85, no. 7 (2004): 1771–89.

Brown, Joseph Epes. *Animals of the Soul: Sacred Animals of the Oglala Sioux*. Rockport, MA: Element, 1992.

Brown, Margaret Kimball. *The Zimmerman Site: Further Excavations at the Grand Village of Kaskaskia*. Reports of Investigations No. 32. Springfield: Illinois State Museum, 1975.

Brown, Nicholas A., and Sarah E. Kanouse. *Re-Collecting Black Hawk: Landscape, Memory, and Power in the American Midwest.* Pittsburgh, PA: University of Pittsburgh Press, 2015.

Buehler, Kent J. "Getting Over the Hump: Some Concluding Remarks on Southern Plains Bison Procurement/Utilization Studies." *Plains Anthropologist* 42, no. 159 (1997): 173.

Buffalo, Jonathan Lantz. "Oral History of the Meskwaki." *Wisconsin Archeologist* 89, no. 1–2 (2008): 3–6.

Buss, James Joseph. *Winning the West with Words: Language and Conquest in the Lower Great Lakes.* Norman: University of Oklahoma Press, 2011.

Cadillac, Antoine Laumet de Lamothe. "Memoir." In *The Western Country in the 17th Century*, Edited by Milo Milton Quaife, 3–8. Chicago: Lakeside, 1947.

Caldwell, Norman W. "Charles Juchereau de St. Denys: A French Pioneer in the Mississippi Valley." *Mississippi Valley Historical Review* 28, no. 4 (1942): 563–80.

Callender, Charles. "Illinois." In *Handbook of North American Indians, Vol. 15, Northeast,* edited by Bruce G. Trigger, 673–80. Washington, DC: Smithsonian Institution, 1978.

Callison, Candis. *How Climate Change Comes to Matter: The Communal Life of Facts.* Durham, NC: Duke University Press, 2014.

Callon, Michel, and John Law. "Agency and the Hybrid Collectif." *South Atlantic Quarterly* 94, no. 2 (1995): 481–507.

Calloway, Colin G. *One Vast Winter Count: The Native American West before Lewis and Clark.* Lincoln: University of Nebraska Press, 2003.

Cameron, Catherine M. "The Effects of Warfare and Captive-Taking on Indigenous Mortality in Post-Contact North America." In *Beyond Germs: Native Depopulation in North America,* edited by Catherine M. Cameron, Paul Kelton, and Alan C. Swedlund, 174–97. Tucson: University of Arizona Press, 2015.

Cameron, Catherine M., Paul Kelton, and Alan C. Swedlund, eds. *Beyond Germs: Native Depopulation in North America.* Tucson: University of Arizona Press, 2015.

Camill, Philip, Charles E. Umbanhowar Jr., Rebecca Teed, Christoph E. Geiss, Jessica Aldinger, Leah Dvorak, Jon Kenning, Jacob Limmer, and Kristina Walkup. "Late-Glacial and Holocene Climatic Effects on Fire and Vegetation Dynamics at the Prairie-Forest Ecotone in South-Central Minnesota." *Journal of Ecology* 91, no. 5 (2003): 822–36.

Carayon, Céline. *Eloquence Embodied: Nonverbal Communication among French and Indigenous Peoples in the Americas.* Chapel Hill: University of North Carolina Press, 2019.

Changnon, Stanley A., James R. Angel, Kenneth E. Kunkel, and Christopher M. B. Lehmann. *Climate Atlas of Illinois.* Champaign: Illinois State Water Survey, 2004.

Changnon, Stanley A., Kenneth E. Kunkel, and Derek Winstanley. "Climate Factors that Caused the Unique Tall Grass Prairie in the Central United States." *Physical Geography* 23, no. 4 (2002): 259–80.

Chaplin, Joyce. *Subject Matter: Technology, the Body, and Science on the Anglo-American Frontier, 1500–1676.* Cambridge, MA: Harvard University Press, 2001.

Charlevoix, Pierre-François-Xavier de. *History and General Description of New France.* 6 vols. Edited by John Gilmary Shea. New York: J. G. Shea, 1866–72.

———. *Journal of a Voyage to North America.* 2 vols. Edited by Louise Phelps Kellogg. Chicago: Caxton, 1923.

Clements, Frederic E. *Research Methods in Ecology.* Lincoln, NE: University Publishing, 1905.

Cobb, Charles R., and Brian M. Butler. "The Vacant Quarter Revisited: Late Mississippian Abandonment of the Lower Ohio Valley." *American Antiquity* 67, no. 4 (October 1, 2002): 625–41.

Collins, Scott L., Alan K. Knapp, John M. Briggs, John M. Blair, and Ernest M. Steinauer. "Modulation of Diversity by Grazing and Mowing in Native Tallgrass Prairie." *Science* 280, no. 5364 (1998): 745–47.

Collins, Scott L., and Linda L. Wallace, eds. *Fire in North American Tallgrass Prairies.* Norman: University of Oklahoma Press, 1990.

Comité de salut public. *Programmes des cours révolutionnaires sur l'art militaire, l'administration militaire, la santé des troupes et les moyens de la conserver.* Paris: Imprimés par ordre du Comité de salut public, 1794.

Cook, E. R., and P. J. Krusic. "North American Drought Atlas." Accessed June 9, 2016. http://iridl.ldeo.columbia.edu.

Cooper, Judith Rose. "Bison Hunting and Late Prehistoric Human Subsistence Economies in the Great Plains." PhD diss., Southern Methodist University, 2008.

Cooter, William. *Ecotones and Broad Spectrum Economies.* Norman: University of Oklahoma Press, 1974.

Costa, David J., and Daryl Baldwin, eds. *Myaamia Neehi Peewaalia Aacimoona Neehi Aalhsoohkaana: Myaamia and Peoria Narratives and Winter Stories.* Myaamia Project of the Miami Tribe of Oklahoma and the Peoria Tribe of Oklahoma, 2010.

Cronon, William. *Changes in the Land: Indians, Colonists, and the Ecology of New England.* New York: Hill & Wang, 1983.

———. *Nature's Metropolis: Chicago and the Great West.* New York: W. W. Norton, 1992.

———. "Paradigm Shift." *Reviews in American History* 11, no. 1 (1983): 93–98.

———. "A Place for Stories: Nature, History, and Narrative." *Journal of American History* 78, no. 4 (1992): 1347–76.

———. "Why Edge Effects?." *Edge Effects* (blog), Center for Culture, History, and Environment. Accessed October 9, 2014. https://edgeeffects.net.

Cronon, William, Jay Gitlin, and George Miles. "Becoming West: Towards a New Meaning for Western History." In *Under an Open Sky: Rethinking America's Western Past,* edited by William Cronon, Jay Gitlin, and George Miles, 3–25. New York: W. W. Norton, 1993.

Crosby, Alfred W. *The Columbian Exchange: Biological and Cultural Consequences of 1492.* Westport, CT: Greenwood, 1972.

———. *Ecological Imperialism: The Biological Expansion of Europe, 900–1900.* New York: Cambridge University Press, 2004.

———. "Virgin Soil Epidemics as a Factor in the Aboriginal Depopulation in America." *William and Mary Quarterly* 33 (1976): 289–99.

Crouch, Christian Ayne. *Nobility Lost: French and Canadian Martial Cultures, Indians, and the End of New France*. Ithaca, NY: Cornell University Press, 2014.

Crumley, Carole L., ed. *Historical Ecology: Cultural Knowledge and Changing Landscapes*. Santa Fe, NM: School for Advanced Research Press, 1994.

D'Artaguiette, Diron. "Journal of Diron D'Artaguiette." In *Travels in the American Colonies*, edited by Newton Dennison Mereness, 16–94. New York: Macmillan, 1916.

DeCoster, Jonathan. "Tangled in the Web: Communication, Power, and the Virgin Soil Hypothesis." *Reviews in American History* 45, no. 4 (2017), 545–51.

DeLay, Brian. *War of a Thousand Deserts: Indian Raids and the U.S.-Mexican War*. New Haven, CT: Yale University Press, 2008.

De León, Jason. *The Land of Open Graves: Living and Dying on the Migrant Trail*. Berkeley: University of California Press, 2015.

Delucia, Christine M. *Memory Lands: King Philip's War and the Place of Violence in the Northeast*. New Haven, CT: Yale University Press, 2018.

DeMallie, Raymond J. "Kinship: The Foundation for Native American Society." In *Studying Native America: Problems and Prospects*, edited by Russell Thornton, 306–56. Madison: University of Wisconsin Press, 1998.

Demuth, Bathsheba. *Floating Coast: An Environmental History of the Bering Strait*. New York: W. W. Norton, 2019.

Denevan, William M. "The Pristine Myth: The Landscape of the Americas in 1492." *Annals of the Association of American Geographers* 82, no. 3 (1992): 369–85.

de Tonti, Henri. "Memoir on LaSalle's Discoveries, By Tonty, 1678–1690." In *Early Narratives of the Northwest, 1634–1699*, edited by Louise Phelps Kellogg, 283–322. New York: Charles Scribner's Sons, 1917.

———. *Relation of Henri de Tonty Concerning the Explorations of La Salle from 1678 to 1683*. Translated by Melville Best Anderson. Chicago: Caxton, 1898.

Diamond, Jared M. *Collapse: How Societies Choose to Fail or Succeed*. New York: Viking, 2005.

Diderot, Denis, and Jean le Rond d'Alembert. *Encyclopédie*. 18 vols. Edited by Robert Morrissey and Glenn Roe. Chicago: University of Chicago/ARTFL Encyclopédie Project, 2017. http://encyclopedie.uchicago.edu/.

Dillehay, Tom D. "Late Quaternary Bison Population Changes on the Southern Plains." *Plains Anthropologist* 19, no. 65 (1974): 180–96.

Drooker, Penelope B., and C. Wesley Cowan. "Transformation of the Fort Ancient Cultures of the Central Ohio Valley." In *Societies in Eclipse: Archaeology of the Eastern Woodlands Indians, AD 1400–1700*, edited by David S. Brose, C. Wesley Cowan, and Robert C. Mainfort, 83–106. Washington, DC: Smithsonian Institution Press, 2001.

Du Chesneau, Jacques. "Memoir on the Western Indians, September 13, 1681." In *Documents Relative to the Colonial History of the State of New York*, edited by Edmund B. O'Callaghan, 9:161–63. Albany: Weed, Parsons, 1853.

Dumont de Montigny. *Mémoires historiques sur la Louisiane*. Paris: C. J. B. Bauche, 1753.

———. *The Memoir of Lieutenant Dumont, 1715–1747: A Sojourner in the French Atlantic*. Edited by Gordon M. Sayre and Carla Zecher. Chapel Hill: University of North Carolina Press, 2012.

Dunbar, Rowland, A. G. Sanders, and Patricia Kay Galloway. *Mississippi Provincial Archives*. 5 vols. Baton Rouge: Louisiana State University Press, 1927–84.

DuVal, Kathleen. "Indian Intermarriage and Métissage in Colonial Louisiana." *William and Mary Quarterly* 65, no. 2 (2008): 267–304.

———. *The Native Ground: Indians and Colonists in the Heart of the Continent*. Philadelphia: University of Pennsylvania Press, 2006.

Dyreson, Mark. "The Foot Runners Conquer Mexico and Texas: Endurance Racing, 'Indigenismo,' and Nationalism." *Journal of Sport History* 31, no. 1 (2004): 1–31.

Edmunds, R. David. *The Potawatomis, Keepers of the Fire*. Norman: University of Oklahoma Press, 1978.

Edmunds, R. David, and Joseph L. Peyser. *The Fox Wars: The Mesquakie Challenge to New France*. Norman: University of Oklahoma Press, 1993.

Edwards, Erika J., Colin P. Osborne, Caroline A. E. Strömberg, Stephen A. Smith, William J. Bond, Pascal-Antoine Christin, Asaph B. Cousins, Melvin R. Duvall, David L. Fox, Robert P. Freckleton, et al. "The Origins of C4 Grasslands: Integrating Evolutionary and Ecosystem Science." *Science* 328, no. 5978 (2010): 587–91.

Edwards, Tai S. *Osage Women and Empire: Gender and Power*. Lawrence: University Press of Kansas, 2018.

Edwards, Tai S., and Paul Kelton. "Germs, Genocides, and America's Indigenous Peoples." *Journal of American History* 107, no. 1 (2020): 52–76.

Egan, Dave, and Evelyn A. Howell. "Introduction." In *The Historical Ecology Handbook: A Restorationist's Guide to Reference Ecosystems*, edited by Dave Egan and Evelyn A. Howell, 1–28. Washington, DC: Island, 2005.

Ellis, Richard N., and Charlie R. Steen. "An Indian Delegation in France, 1725." *Journal of the Illinois State Historical Society* 67 (1974): 385–405.

Emerson, Thomas E., and James Allison Brown. "The Late Prehistory and Protohistory of Illinois." In *Calumet and Fleur-de-Lys: Archaeology of French and Indian Contact in the Mid-Continent*, edited by John A. Walthall and Thomas Emerson, 77–128. Washington, DC: Smithsonian Institution, 1992.

Emerson, Thomas E., and Kristin Hedman. "The Dangers of Diversity: The Consolidation and Dissolution of Cahokia, Native North America's First Urban Polity." In *Beyond Collapse: Archaeological Perspectives on Resilience, Revitalization, and Transformation in Complex Societies*, edited by Ronald K. Faulseit, 147–75. Carbondale: Southern Illinois University Press, 2016.

Emerson, Thomas E., and R. Barry Lewis. *Cahokia and the Hinterlands: Middle Mississippian Cultures of the Midwest*. Urbana: University of Illinois Press, 2000.

Esarey, Duane, and Lawrence A. Conrad. "The Bold Counselor Phase of the Central Illinois River Valley: Oneota's Middle Mississippian Margin." *Wisconsin Archeologist* 79, no. 2 (1998): 38–61.

Ethridge, Robbie. *From Chicaza to Chickasaw: The European Invasion and the Transformation of the Mississippian World, 1540–1715*. Chapel Hill: University of North Carolina Press, 2010.

Faribault-Beauregard, Marthe. *La population des forts Français d'Amérique (XVIIIe siecle): répertoire des baptêmes, mariages et sépultures célébrés dans les forts et les établissements français en Amérique du Nord au XVIIIe siècle*. Montreal: Bergeron, 1982.

Farmer, Jared. "Borderlands of Brutality." *Reviews in American History* 37, no. 4 (2009): 544–52.

Faulkner, Charles H. *The Late Prehistoric Occupation of Northwestern Indiana: A Study of the Upper Mississippi Cultures of the Kankakee Valley*. Prehistory Research Series vol. 5, no. 1. Indianapolis: Indiana Historical Society, 1972.

Faulseit, Ronald K., J. Heath Anderson, Christina Conley, Stacy Dunn, Jerry Ek, and Thomas Emerson. "Collapse, Resilience, and Transformation in Complex Societies: Modeling Trends and Understanding Diversity." In *Beyond Collapse: Archaeological Perspectives on Resilience, Revitalization, and Transformation in Complex Societies*, edited by Ronald K. Faulseit, 24–47. Carbondale: Southern Illinois University Press, 2016.

Fauteux, Joseph Noël. *Essai sur l'industrie au Canada sous le régime français*. 2 vols. Québec: Ls-A. Proulx, 1927.

Fei, Songlin, Johanna M. Desprez, Kevin M. Potter, Insu Jo, Jonathan A. Knott, and Christopher M. Oswalt. "Divergence of Species Responses to Climate Change." *Science Advances* 3, no. 5 (2017).

Fenn, Elizabeth A. *Encounters at the Heart of the World: A History of the Mandan People*. New York: Hill and Wang, 2014.

Flores, Dan. *American Serengeti: The Last Big Animals of the Great Plains*. Lawrence: University Press of Kansas, 2016.

———. "Bison Ecology and Bison Diplomacy: The Southern Plains from 1800 to 1850." *Journal of American History* 78, no. 2 (1991): 465–85.

———. *Caprock Canyonlands: Journeys into the Heart of the Southern Plains*. Austin: University of Texas Press, 1990.

———. "Place: An Argument for Bioregional History." *Environmental History Review* 18, no. 4 (1994): 1–18.

Forbis, Richard G. "Some Facets of Communal Hunting." *Plains Anthropologist* 23, no. 82 (1978): 3–8.

Foreman, Dave. "The Myth of the Humanized Pre-Columbian Landscape." In *Keeping the Wild: Against the Domestication of Earth*, edited by George Wuerthner, Eileen Crist, and Tom Butler, 114–25. Washington, DC: Island, 2014.

Fortier, John, and Donald Chaput. "A Historical Reexamination of Juchereau's Illinois Tannery." *Journal of the Illinois State Historical Society* 62, no. 4 (1969): 385–406.

Frink, Lisa, and Kathryn Weedman, eds. *Gender and Hide Production*. Walnut Creek, CA: AltaMira, 2005.

Gallay, Alan. *The Indian Slave Trade: The Rise of the English Empire in the American South, 1670–1717*. New Haven, CT: Yale University Press, 2002.

Galloway, Patricia Kay. *Choctaw Genesis, 1500–1700*. Lincoln: University of Nebraska Press, 1995.

Gibbon, Guy E. "Cultural Dynamics and the Development of the Oneota Life-Way in Wisconsin." *American Antiquity* 37, no. 2 (1972): 166–85.

———. "Oneota." In *Encyclopedia of Prehistory, Volume 6: North America*, edited by Peter N. Peregrine and Melvin Ember, 389–407. Boston: Springer, 2001.

Gilbert, Matthew Sakiestewa. "Hopi Footraces and American Marathons, 1912–1930." *American Quarterly* 62, no. 1 (2010): 77–101.

———. *Hopi Runners: Crossing the Terrain between Indian and American*. Lawrence: University Press of Kansas, 2018.

Gleason, Henry Allan. "The Relation of Forest Distribution and Prairie Fires in the Middle West." *Torreya* 13, no. 8 (1913): 173–81.

———. "The Vegetational History of the Middle West." *Annals of the Association of American Geographers* 12 (1922): 39–85.

Gonella, Michael Paul. "Myaamia Ethnobotany." PhD diss., Miami University, 2007.

Goodman, Alan H., and George J. Armelagos. "Disease and Death at Dr. Dickson's Mounds." *Natural History* 94, no. 9 (1985): 12–18.

Gosz, James R. "Ecological Functions in a Biome Transition Zone: Translating Local Responses to Broad-Scale Dynamics." In *Landscape Boundaries: Consequences for Biotic Diversity and Ecological Flows*, edited by Andrew J. Hansen and Francesco DiCastri, 55–76. Boston: Springer-Verlag, 1992.

Governanti, Brett J. "The Myaamia Mapping Project." Master's thesis, Miami University, 2005.

Green, William. "Examining Protohistoric Depopulation in the Upper Midwest." *Wisconsin Archeologist* 74, no. 1–4 (1993): 290–323.

Green, William, and David W Benn. *Oneota Archaeology: Past, Present, and Future*. Iowa City: University of Iowa Press, 1995.

Greene, Ann Norton. *Horses at Work: Harnessing Power in Industrial America*. Cambridge, MA: Harvard University Press, 2008.

Greer, Diane, Diane Szafoni, Liane Suloway, Kate Hunter, Janet Jarvis, Ian Crelling, Alissa Eienstein, et al. "Land Cover of Illinois in the Early 1800s." Champaign: Illinois Natural History Survey, 1990.

Griffin, James B. "Climatic Change: A Contributory Cause of the Growth and Decline of Northern Hopewellian Culture." *Wisconsin Archeologist* 41, no. 2 (1960), 21–33.

———. *The Fort Ancient Aspect: Its Cultural and Chronological Position in Mississippi Valley Archaeology*. Ann Arbor: University of Michigan Press, 1943.

———. "A Hypothesis for the Prehistory of the Winnebago." In *Culture in History*, edited by Stanley Diamond, 809–65. New York: Columbia University Press, 1960.

Griffin, John W., and Donald E. Wray. "Bison in Illinois Archaeology." *Transactions of the Illinois State Academy of Science* 38 (1945): 21–26.

Grimm, Eric C. "Chronology and Dynamics of Vegetation Change in the Prairie-Woodland Region of Southern Minnesota, U.S.A." *New Phytologist* 93, no. 2 (1983): 311–50.

Grimm, Eric C., and George L. Jacobson. "Late-Quaternary Vegetation History of the Eastern United States." *Developments in Quaternary Sciences*, no. 1 (2003): 381–402.

Grobbel, William B. "Tanning and Tanneries: An Introduction to Archaeology of the Leather Industry." *North American Archaeologist* 18, no. 2 (October 1, 1997): 121–47.

Gross, Richard, and Craig P. Howard. "Colbert, La Salle, and the Search for Empire." *Journal of the Illinois State Historical Society* 113, no. 2 (2020): 68–101.

Gundersen, James Novotny. "'Catlinite' and the Spread of the Calumet Ceremony." *American Antiquity* 58, no. 3 (1993): 560–62.

Habicht-Mauche, J. A. "The Shifting Role of Women and Women's Labor on the Protohistoric Southern High Plains." In *Gender and Hide Production*, edited by Lisa Frink and Kathryn Weedman, 37–57. Walnut Creek, CA: AltaMira, 2005.

———. "Women on the Edge: Looking at Protohistoric Plains-Pueblo Interaction from a Feminist Perspective." In *The Oxford Handbook of North American Archaeology*, edited by Tim Pauketat, 386–97. New York: Oxford University Press, 2012.

Hall, Robert L. *An Archaeology of the Soul: North American Indian Belief and Ritual.* Urbana: University of Illinois Press, 1997.

Hämäläinen, Pekka. "The Changing Histories of North America before Europeans." *OAH Magazine of History* 27, no. 4 (2013): 5–7.

———. *The Comanche Empire.* New Haven, CT: Yale University Press, 2009.

———. *Lakota America: A New History of Indigenous Power.* New Haven, CT: Yale University Press, 2019.

———. "The Politics of Grass: European Expansion, Ecological Change, and Indigenous Power in the Southwest Borderlands." *William and Mary Quarterly* 67, no. 2 (2010): 173–208.

———. "The Rise and Fall of Plains Indian Horse Cultures." *Journal of American History* 90, no. 3 (2003): 833–62.

Hämäläinen, Pekka, and Samuel Truett. "On Borderlands." *Journal of American History* 98, no. 2 (2011): 338–61.

Hamilton, W. D. "The Genetical Evolution of Social Behaviour, 1 and 2." *Journal of Theoretical Biology* 7, no. 1 (1964): 1–16, 17–52.

Hanson, Jeffery R. "Bison Ecology in the Northern Plains." *Plains Anthropologist* 29, no. 104 (1984): 93–113.

Harn, Alan, and Terrance J. Martin. "Early Confrontations with the Illinois Country's 'Wild Cattle.'" *Living Museum* 67, no. 4 (2005): 9–14.

Harrod, Howard L. *The Animals Came Dancing: Native American Sacred Ecology and Animal Kinship.* Tucson: University of Arizona Press, 2000.

Hauser, Raymond E. "The Berdache and the Illinois Indian Tribe during the Last Half of the Seventeenth Century." *Ethnohistory* 37, no. 1 (January 1, 1990): 45–65.

———. "The Illinois Indian Tribe: From Autonomy and Self-Sufficiency to Dependency and Depopulation." *Journal of the Illinois State Historical Society* 69, no. 2 (May 1976): 127–38.

———. "Warfare and the Illinois Indian Tribe During the Seventeenth Century: An Exercise in Ethnohistory." *Old Northwest* 10, no. 4 (1984): 367–87.

Havard, Gilles. *Empire et métissages: Indiens et Français dans le pays d'en haut, 1660–1715*. Sillery, QC: Septentrion, 2003.

———. *The Great Peace of Montreal of 1701: French-Native Diplomacy in the 17th Century*. Montreal: McGill-Queen's University Press, 2001.

———. "Un Américain à Rochefort (1731–1732): Le destin de Coulipa, Indien renard." In *Les étrangers dans les villes-ports atlantiques: expériences françaises et allemandes (XVe–XIXe siècle)*, edited by Mickaël Augeron, 143–55. Paris: Les Indes savantes, 2010.

Hecht, Susanna B. "Domestication, Domesticated Landscapes, and Tropical Natures." In *The Routledge Companion to the Environmental Humanities*, edited by Ursula K. Heise, Jon Christensen, and Michelle Niemann, 21–34. New York: Routledge, 2017.

Heimmermann, Daniel. "Crisis and Protest in the Guilds of Eighteenth Century France: The Example of the Bordeaux Leather Trades." *Proceedings of the Western Society for French History* 23 (1996): 431–41.

Hennepin, Louis. *A Description of Louisiana*. Edited by John Gilmary Shea. New York: John G. Shea, 1880.

———. *A New Discovery of a Vast Country in America, by Father Louis Hennepin. Reprinted from the Second London Issue of 1698*. 2 vols. Edited by Reuben Gold Thwaites. Chicago: A. C. McClurg, 1903.

Henning, Dale R. "Cultural Adaptations to the Prairie Environment: The Ioway Example." In *Proceedings of the Twelfth North American Prairie Conference: Recapturing a Vanishing Heritage*, edited by Carol A. Jacobs and Daryl Smith, 193–94. Cedar Falls: University of Northern Iowa, 1992.

———. "The Oneota Tradition." In *Archaeology on the Great Plains*, edited by W. Raymond Wood, 345–414. Lawrence: University Press of Kansas, 1998.

Hodge, Adam R. *Ecology and Ethnogenesis: An Environmental History of the Wind River Shoshones, 1000–1868*. Lincoln: University of Nebraska Press, 2019.

Hoganson, Kristin L. *The Heartland: An American History*. New York: Penguin, 2019.

Holm, Tom. "American Indian Warfare: The Cycles of Conflict and the Militarization of Native North America." In *A Companion to American Indian History*, edited by Philip J. Deloria and Neal Salisbury, 154–74. New York: Blackwell, 2007.

Huebner, Jeffrey A. "Late Prehistoric Bison Populations in Central and Southern Texas." *Plains Anthropologist* 36, no. 137 (1991): 343–58.

Hupy, Christina M., and Catherine H. Yansa. "Late Holocene Vegetation History of the Forest Tension Zone in Central Lower Michigan, USA." *Physical Geography* 30, no. 3 (May 1, 2009): 205–35.

ILDA (Indigenous Language Digital Archive). Miami-Illinois, created by Myaamia Center, Oxford, Ohio. https://mc.miamioh.edu/ilda-myaamia/about.

Ingold, Tim. *The Perception of the Environment: Essays on Livelihood, Dwelling, and Skill*. New York: Routledge, 2000.

Ironstrack, George. "The Good Path: Part 1." *Aacimotaatiiyankwi* (blog), November 1, 2011. https://aacimotaatiiyankwi.org/2011/11/01/the-good-path-part-i/.

Ironstrack, George, Liz Ellis, Cam Shriver, Scott Shoemaker, Dave Costa, and Bob Morrissey. "Ciinkwia Minohsaya 'Painted Thunderbird Robe' Series." *Aacimotaatiiyankwi* (blog), December 15, 2020. https://aacimotaatiiyankwi.org/2020/12/15/ciinkwia-minohsaya-painted-thunderbird-robe-series/.

Isenberg, Andrew C. *The Destruction of the Bison: An Environmental History, 1750–1920*. New York: Cambridge University Press, 2000.

———. "Seas of Grass: Grasslands in World Environmental History." In *The Oxford Handbook of Environmental History*, edited by Andrew C. Isenberg, 149–53. New York: Oxford University Press, 2014.

ISHL (Illinois State Historical Library). *Collections of the Illinois State Historical Library*. 38 vols. Illinois State Historical Library, Springfield, 1903–70.

Jablow, Joseph. *The Cheyenne in Plains Indian Trade Relations, 1795–1840*. Lincoln: University of Nebraska Press, 1994.

Jacoby, Karl. *Shadows at Dawn: A Borderlands Massacre and the Violence of History*. New York: Penguin, 2008.

Jakle, John A. "The American Bison and the Human Occupance of the Ohio Valley." *Proceedings of the American Philosophical Society* 112, no. 4 (August 15, 1968): 299–305.

Johnson, Walter. "On Agency." *Journal of Social History* 37, no. 1 (2003): 113–24.

Jones, Christopher F. "Petromyopia: Oil and the Energy Humanities." *Humanities* 5, no. 2 (June 2016): 36–46.

Jones, David S. "Virgin Soils Revisited." *William and Mary Quarterly* 60, no. 4 (October 1, 2003): 703–42.

Jones, Linda Carol. *The Shattered Cross: French Catholic Missionaries on the Mississippi River, 1698–1725*. Baton Rouge: Louisiana State University Press, 2020.

Jones, William. *Fox Texts: Publications of the American Ethnological Society*. Leyden: Brill, 1907.

Jordan, Terry G. "Between the Forest and the Prairie." *Agricultural History* 38, no. 4 (1964): 205–16.

Kehoe, Alice Beck. "Cahokia, the Great City." *OAH Magazine of History* 27, no. 4 (2013): 17–21.

———. *North America before the European Invasions*. New York: Routledge, 2017.

Kellogg, Elizabeth A. "Evolutionary History of the Grasses." *Plant Physiology* 125, no. 3 (2001): 1198–1205.

Kellogg, Louise Phelps, ed. *Early Narratives of the Northwest, 1634–1699*. New York: Scribner's Sons, 1917.

———. *The French Régime in Wisconsin and the Northwest*. Madison: State Historical Society of Wisconsin, 1925.

Kelton, Paul K. *Epidemics and Enslavement: Biological Catastrophe in the Native Southeast, 1492–1715*. Lincoln: University of Nebraska Press, 2007.

King, James E. "Late Quaternary Vegetational History of Illinois." *Ecological Monographs* 51, no. 1 (1981): 43–62.

———. "Post-Pleistocene Vegetational Changes in the Midwestern United States." In *Archaic Prehistory on the Prairie-Plains Border*, edited by Alfred E. Johnson, 3–11. Lawrence: University Press of Kansas, 1980.

———. "The Prairies of Illinois." Prairie Preservation Society of Ogle County, Illinois, March 1982.

Kluckhohn, Clyde. "Skin Dressing." In *Navaho Material Culture*. Cambridge, MA: Belknap Press of Harvard University Press, 1971.

Knapp, Alan K., John M. Blair, John M. Briggs, Scott L. Collins, David C. Hartnett, Loretta C. Johnson, and E. Gene Towne. "The Keystone Role of Bison in North American Tallgrass Prairie." *BioScience* 49, no. 1 (1999): 39–50.

Knapp, Alan K., and Ernesto Medina. "Success of C4 Photosynthesis in the Field: Lessons from Communities Dominated by C4 Plants." In *C4 Plant Biology*, edited by Rowan F. Sage and Russell K. Monson, 251–83. San Diego: Academic, 1999.

Kolata, Dennis R., and Cheryl Nimz, eds. *Geology of Illinois*. Champaign: Illinois State Geological Survey, 2010.

Krech, Shepard, III. *The Ecological Indian: Myth and History*. New York: W. W. Norton, 1999.

Krutak, Lars F. *Tattoo Traditions of Native North America: Ancient and Contemporary Expressions of Identity*. Arnhem, Netherlands: LM, 2014.

Küchler, A. W. "Potential Natural Vegetation (United States)" map. Washington, DC: U.S. Geological Survey, 1985.

———. "Problems in Classifying and Mapping Vegetation for Ecological Regionalization." *Ecology* 54, no. 3 (1973): 512–23.

Lahontan, Louis Armand de Lom d'Arce. *New Voyages to North-America*. Edited by Reuben Gold Thwaites and Victor Hugo Paltsits. Chicago: A. C. McClurg, 1905.

Lallement. "A Copy of a Letter of Lallemant to the Directors of the Company of the Indies, Dated from Caskaskias, April 5, 1721." Kaskaskia Papers, Chicago History Museum.

La Potherie, Claude Charles Le Roy. "History of the Savage Peoples Who Are Allies of New France." In *The Indian Tribes of the Upper Mississippi Valley and Region of the Great Lakes*, edited by Emma Helen Blair, 1:273–372, 2:13–138. Lincoln: University of Nebraska Press, 1996.

La Salle, Nicolas de. *Relation of the Discovery of the Mississippi River*. Edited by Melville Best Anderson. Chicago: Caxton, 1898.

La Salle, Robert Cavelier. *Relation of the Discoveries and Voyages of Cavelier De La Salle from 1679 to 1681, the Official Narrative*. Edited by Melville Best Anderson. Chicago: Caxton, 1901.

La Source, Dominic Thaumur de. "Voyage Up the Mississippi in 1699–1700." In *Early Voyages Up and Down the Mississippi*, edited by John Gilmary Shea, 79–86. Albany: J. Munsell, 1861.

Lasteyrie, Charles-Philibert de. *Histoire naturelle et économique du boeuf, de la vache et du Buffle . . . Par M. C.-P. de Lasteyrie*. Self-published, 1834.

Latour, Bruno. *Reassembling the Social: An Introduction to Actor-Network-Theory.* New York: Oxford University Press, 2005.

Lauck, Jon. *The Lost Region: Toward a Revival of Midwestern History.* Iowa City: University of Iowa Press, 2013.

Law, John. "Technology and Heterogeneous Engineering: The Case of Portuguese Expansion." In *The Social Construction of Technological Systems: New Directions in the Sociology and History of Technology,* edited by Wiebe E. Bijker, Thomas P. Hughes, and Trevor Pinch, 105–27. Cambridge: MIT Press, 2012.

LeCain, Timothy J. *The Matter of History: How Things Create the Past.* New York: Cambridge University Press, 2017.

Lee, Jacob F. *Masters of the Middle Waters: Indian Nations and Colonial Ambitions along the Mississippi.* Cambridge, MA: Harvard University Press, 2019.

Lehmer, Donald J. "The Plains Bison Hunt—Prehistoric and Historic." *Plains Anthropologist* 8, no. 22 (1963): 211–17.

LeMenager, Stephanie. *Living Oil: Petroleum Culture in the American Century.* New York: Oxford University Press, 2014.

Leopold, Aldo. *Game Management.* Madison: University of Wisconsin Press, 1986.

———. *A Sand County Almanac.* New York: Sierra Club/Ballatine, 1970.

Lepore, Jill. *The Name of War: King Philip's War and the Origins of American Identity.* New York: Knopf , 1998.

Liette, Pierre de. "Memoir Concerning the Illinois Country, ca. 1693." In *The French Foundations,* edited by Theodore Calvin Pease and Raymond C. Werner, 23:302–95. Collections of the Illinois State Historical Library, Springfield, 1934.

Lopinot, Neal, and William Woods. "Wood Overexploitation and the Collapse of Cahokia." In *Foraging and Farming in the Eastern Woodlands,* edited by C. Margaret Scarry, 206–31. Gainesville: University Press of Florida, 1993.

Lott, Dale F. *American Bison: A Natural History.* Berkeley: University of California Press, 2002.

Louis XV. *Ordonnance du Roi, concernant la cavalerie.* Paris, 1763.

Macfarlane, Robert. *The Old Ways: A Journey on Foot.* New York: Viking, 2012.

Madson, John. *Where the Sky Began: Land of the Tallgrass Prairie.* Iowa City: University of Iowa Press, 2004.

Mann, Charles C. *1491: New Revelations of the Americas before Columbus.* New York: Knopf, 2005.

———. *1493: Uncovering the New World Columbus Created.* New York: Knopf, 2011.

Margry, Pierre. *Découvertes et établissements des Français dans l'ouest et dans le sud de l'Amérique Septentrionale (1614–1754).* 6 vols. Paris: D. Jouaust, 1879.

Martin, Terrance J., and Alan Harn. "The Lonza-Caterpillar Site: Bison Bone Deposits from the Illinois River, Peoria County, Illinois." In *Records of Early Bison in Illinois,* edited by R. Bruce McMillan, 19–66. Springfield: Illinois State Museum, 2006.

Mason, Otis Tufton. *Aboriginal Skin-Dressing: A Study Based on Material in the U. S. National Museum.* Washington, DC, 1891.

Mason, Ronald J. "Oneota and Winnebago Ethnogenesis: An Overview." *Wisconsin Archaeologist* 74 (March-December 1993): 400–422.

Mazrim, Robert. "The Danner Series Pottery of the Illinois." In *Protohistory at the Grand Village of the Kaskaskia: The Illinois Country on the Eve of Colony*, edited by Robert Mazrim, 29–50. Urbana: Prairie Research Institute, 2015.

Mazrim, Robert, and Duane Esarey. "Rethinking the Dawn of History: The Schedule, Signature, and Agency of European Goods in Protohistoric Illinois." *Midcontinental Journal of Archaeology* 32, no. 2 (October 1, 2007): 145–200.

McCafferty, Michael. "The Illinois Place Name 'Pimitéoui.'" *Journal of the Illinois State Historical Society* 102, no. 2 (July 1, 2009): 177–92.

———. "Illinois Voices," In *Protohistory at the Grand Village of the Kaskaskia: The Illinois Country on the Eve of Colony*, edited by Robert Mazrim, 117–30. Urbana: Prairie Research Institute, 2015.

McCann, Joseph. "Before 1492: The Making of the Pre-Columbian Landscape: Part 1 and 2." *Ecological Restoration, North America* 17, nos. 1–2, 3 (1999): 15–31, 107–19.

McClain, William E., Charles M. Ruffner, John E. Ebinger, and Greg Spyreas. "Patterns of Anthropogenic Fire within the Midwestern Tallgrass Prairie, 1673–1905: Evidence from Written Accounts." *Natural Areas Journal* 41, no. 4 (2021): 283–300.

McCoy, Tim, George Michael Ironstrack, Daryl Baldwin, Andrew J. Strack, and Wayne Olm. *Asiihkiwi Neehi Kiisikwi Myaamionki: Earth and Sky, the Place of the Myaamiaki*. Miami, OK: Miami Tribe of Oklahoma, 2011.

McDonnell, Michael A. *Masters of Empire: Great Lakes Indians and the Making of America*. New York: Hill and Wang, 2015.

McDougall, Christopher. *Born to Run: A Hidden Tribe, Superathletes, and the Greatest Race the World Has Never Seen*. New York: Knopf, 2009.

McElrath, Dale L., and Thomas E. Emerson. "Concluding Thoughts on the Archaic Occupation of the Eastern Woodlands." In *Archaic Societies: Diversity and Complexity across the Midcontinent*, edited by Thomas E. Emerson, Dale L. McElrath, and Andrew C. Fortier, 841–55. Albany: SUNY Press, 2012.

McMillan, R Bruce. "Bison in Missouri Archaeology." *Missouri Archaeologist* 73 (2012): 79–136.

———. "Perspectives on the Biogeography and Archaeology of Bison in Illinois." In *Records of Early Bison in Illinois*, edited by R. Bruce McMillan, 67–147. Springfield: Illinois State Museum, 2006.

———, ed. *Records of Early Bison in Illinois*. Vol. 31 of Scientific Papers of the Illinois State Museum. Springfield: Illinois State Museum, 2006.

McNeill, John Robert. *The Great Acceleration: An Environmental History of the Anthropocene since 1945*. Cambridge, MA: Harvard University Press, 2014.

McNeill, William Hardy. *Plagues and Peoples*. Garden City, NY: Anchor, 1976.

Meine, Curt. "Foreword." In *The Historical Ecology Handbook: A Restorationist's Guide to Reference Ecosystems*, edited by Dave Egan and Evelyn A. Howell, xv–xix. Washington, DC: Island, 2005.

Merrell, James H. *The Indians' New World: Catawbas and Their Neighbors from European Contact through the Era of Removal*. Chapel Hill: University of North Carolina Press, 1989.

Michalik, Laura K. "An Ecological Perspective on the Huber Phase Subsistence Settlement System." In *Oneota Studies*, edited by Guy E. Gibbon, 29–53. Minneapolis: University of Minnesota Press, 1982.

Michaud, Claudine. "Manufacture de buffles à Corbeil-Essonnes." *Le Papyvore* 29 (2009): 6–9.

Michelson, Truman. "Contributions to Fox Ethnology: Notes on the Ceremonial Runners of the Fox Indians." *Bulletin of the Bureau of American Ethnology*, 1927.

Milner, George R. "Population Decline and Culture Change in the American Midcontinent: Bridging the Prehistoric and Historic Divide." In *Beyond Germs: Native Depopulation in North America*, edited by Catherine M. Cameron, Paul Kelton, and Alan C. Swedlund, 50–73. Tucson: University of Arizona Press, 2015.

Milner, George R., Eve Anderson, and Virginia G. Smith. "Warfare in Late Prehistoric West-Central Illinois." *American Antiquity* 56, no. 4 (1991): 581–603.

Morgan, M. J. *Land of Big Rivers: French and Indian Illinois, 1699–1778*. Carbondale: Southern Illinois University Press, 2010.

Morris, Christopher. "How to Prepare Buffalo, and Other Things the French Taught Indians about Nature." In *French Colonial Louisiana and the Atlantic World*, edited by Bradley G. Bond, 22–42. Baton Rouge: Louisiana State University Press, 2005.

Morris, J. L., J. R. Mueller, A. Nurse, C. J. Long, and K. K. Mclauchlan. "Holocene Fire Regimes, Vegetation and Biogeochemistry of an Ecotone Site in the Great Lakes Region of North America." *Journal of Vegetation Science* 25, no. 6 (2014): 1450–64.

Morrissey, Robert Michael. "Climate, Ecology and History in North America's Tallgrass Prairie Borderlands." *Past & Present* 245, no. 1 (2019): 39–77.

———. "'The Country Is Greatly Injured': Human-Animal Relationships, Ecology and the Fate of Empire in the Eighteenth Century Mississippi Valley Borderlands." *Environment & History* 22, no. 2 (2016): 157–90.

———. "The Power of the Ecotone: Bison, Slavery, and the Rise and Fall of the Grand Village of the Kaskaskia." *Journal of American History* 102, no. 3 (2015): 667–92.

Muñoz, Samuel E. "Forests, Fields, and Floods: A Historical Ecology of the Cahokia Region, Illinois, USA." PhD diss., University of Wisconsin, Madison, 2015.

Muñoz, Samuel E., Kristine E. Gruley, Ashtin Massie, David A. Fike, Sissel Schroeder, and John W. Williams. "Cahokia's Emergence and Decline Coincided with Shifts of Flood Frequency on the Mississippi River." *Proceedings of the National Academy of Sciences of the United States of America* 112, no. 20 (2015): 6319–24.

Muñoz, Samuel E., David J. Mladenoff, Sissel Schroeder, and John W. Williams. "Defining the Spatial Patterns of Historical Land Use Associated with the Indigenous Societies of Eastern North America." *Journal of Biogeography* 41, no. 12 (2014): 2195–2210.

Nabokov, Peter. *Indian Running.* Santa Barbara, CA: Capra, 1981.
Nelson, David M., and Feng Sheng Hu. "Patterns and Drivers of Holocene Vegetational Change Near the Prairie-Forest Ecotone in Minnesota: Revisiting McAndrews' Transect." *New Phytologist* 179, no. 2 (2008): 449–59.
Nelson, David M., Feng Sheng Hu, Eric C. Grimm, B. Brandon Curry, and Jennifer E. Slate. "The Influence of Aridity and Fire on Holocene Prairie Communities in the Eastern Prairie Peninsula." *Ecology* 87, no. 10 (2006): 2523–36.
Nelson, David M., Feng Sheng Hu, Jian Tian, Ivanka Stefanova, and Thomas A. Brown. "Response of C3 and C4 Plants to Middle-Holocene Climatic Variation near the Prairie-Forest Ecotone of Minnesota." *Proceedings of the National Academy of Sciences of the United States of America* 101, no. 2 (2004): 562–67.
Norall, Frank. *Bourgmont: Explorer of the Missouri, 1698–1725.* Lincoln: University of Nebraska Press, 1988.
O'Brien, Jean M. *Firsting and Lasting: Writing Indians Out of Existence in New England.* Minneapolis: University of Minnesota Press, 2010.
O'Gorman, Emily, and Andrea Gaynor. "More-Than-Human Histories." *Environmental History* 25, no. 4 (2020): 711–35.
Oliver, Chad. *Ecology and Cultural Continuity as Contributing Factors in the Social Organization of the Plains Indians.* Berkeley: University of California Press, 1962.
Osborne, Colin P., and David J. Beerling. "Nature's Green Revolution: The Remarkable Evolutionary Rise of C4 Plants." *Philosophical Transactions of the Royal Society B: Biological Sciences* 361, no. 1465 (2006): 173–94.
Ostler, Jeffrey. *Surviving Genocide: Native Nations and the United States from the American Revolution to Bleeding Kansas.* New Haven, CT: Yale University Press, 2019.
Ostler, Jeffrey, and Nancy Shoemaker. "Settler Colonialism in Early American History: Introduction." *William and Mary Quarterly* 76, no. 3 (2019): 361–68.
Palm, Mary Borgias. *The Jesuit Missions of the Illinois Country, 1673–1763.* PhD diss., St. Louis University, 1931.
Parmalee, Paul W. "Cave and Archaeological Faunal Deposits as Indicators of Post-Pleistocene Animal Populations and Distribution in Illinois." *Quaternary of Illinois* 14 (1968): 104–13.
———. "The Faunal Complex of the Fisher Site, Illinois." *American Midland Naturalist* 68, no. 2 (1962): 399–408.
Parmenter, John. *The Edge of the Woods: Iroquoia, 1534–1701.* East Lansing: Michigan State University Press, 2010.
Pénicaut, André. *Fleur de Lys and Calumet Being the Pénicaut Narrative of French Adventure in Louisiana.* Edited by Richebourg Gaillard McWilliams. Baton Rouge: Louisiana State University Press, 1953.
Perrot, Nicolas. "Memoir on the Manners, Customs, and Religion of the Savages of North America." In *The Indian Tribes of the Upper Mississippi Valley and Region of the Great Lakes,* edited by Emma Helen Blair, 1:22–272. Lincoln: University of Nebraska Press, 1996.

Pesantubbee, Michelene E. *Choctaw Women in a Chaotic World: The Clash of Cultures in the Colonial Southeast.* Albuquerque: University of New Mexico Press, 2005.

Peyser, Joseph L. "The Fate of the Fox Survivors: A Dark Chapter in the History of the French in the Upper Country, 1726–1737." *Wisconsin Magazine of History* 73, no. 2 (1989): 83–110.

———. "The 1730 Siege of the Foxes: Two Maps by Canadian Participants Provide Additional Information on the Fort and Its Location." *Illinois Historical Journal* 80, no. 3 (October 1, 1987): 147–54.

Pianka, Eric R. *Evolutionary Ecology.* New York: Benjamin Cummings, 2000.

Pielou, E. C. *After the Ice Age: The Return of Life to Glaciated North America.* Chicago: University of Chicago Press, 1991.

Pollan, Michael. *The Omnivore's Dilemma: A Natural History of Four Meals.* New York: Penguin, 2007.

Pond, Peter. "Journal of Peter Pond, 1740–1755." In *Wisconsin Historical Collections.* Vol. 18. Madison: State Historical Society of Wisconsin, 1908.

Pratz, Le Page du. *The History of Louisiana, or of the Western Parts of Virginia and Carolina.* 2 vols. London: T. Becket and P. A. De Hondt, 1763.

Prince, Hugh C. *Wetlands of the American Midwest: A Historical Geography of Changing Attitudes.* Chicago: University of Chicago Press, 1997.

Pyne, Stephen J. *Fire: A Brief History.* Seattle: University of Washington Press, 2001.

Raudot, Antoine Denis. "Memoir Concerning the Different Indian Nations of North America." In *The Indians of the Western Great Lakes, 1615–1760*, edited by W. Vernon Kinietz, 341–410. Ann Arbor: University of Michigan Press, 1940.

Rhoades, Robert E. "The Ecotone Concept in Anthropology and Ecology." *Papers in Anthropology (University of Oklahoma)* 15 (1974): 23–36.

Rice, James D. *Nature and History in the Potomac Country: From Hunter-Gatherers to the Age of Jefferson.* Baltimore: Johns Hopkins University Press, 2009.

———. "War and Politics: Powhatan Expansionism and the Problem of Native American Warfare." *William and Mary Quarterly* 77, no. 1 (2020): 3–32.

Richter, Daniel K. *Before the Revolution: America's Ancient Pasts.* Cambridge, MA: Belknap Press of Harvard University Press, 2011.

———. *Facing East from Indian Country: A Native History of Early America.* Cambridge, MA: Harvard University Press, 2003.

———. *The Ordeal of the Longhouse: The Peoples of the Iroquois League in the Era of European Colonization.* Chapel Hill: University of North Carolina Press, 1992.

———. "War and Culture: The Iroquois Experience." *William and Mary Quarterly* 40, no. 4 (1983): 528–59.

Riello, Giorgio. "Nature, Production and Regulation in Eighteenth-Century Britain and France: The Case of the Leather Industry." *Historical Research* 81, no. 211 (February 2008): 75–99.

Rindos, David, and Sissel Johannessen. "Human-Plant Interactions and Cultural Change in the American Bottom." In *Cahokia and the Hinterlands: Middle Mississippian Cultures of the Midwest*, edited by Thomas E. Emerson and R. Barry Lewis, 35–45. Springfield: Illinois Historic Preservation Agency, 1991.

Risser, Paul G. "The Status of the Science Examining Ecotones." *BioScience* 45, no. 5 (1995): 318–25.
———, ed. *The True Prairie Ecosystem*. Stroudsburg, PA: Hutchinson Ross, 1981.
Ritterbush, Lauren W. "Drawn by the Bison: Late Prehistoric Native Migration into the Central Plains." *Great Plains Quarterly* 22, no. 4 (2002): 259–70.
Robbins, Paul. *Lawn People: How Grasses, Weeds, and Chemicals Make Us Who We Are*. Philadelphia: Temple University Press, 2007.
Robertson, Kenneth R., Roger C. Anderson, and Mark W. Schwartz. "The Tallgrass Prairie Mosaic." In *Conservation in Highly Fragmented Landscapes*, edited by Mark W. Schwartz, 55–87. Boston: Springer, 1997.
Rodgers, Cassandra S., and Roger C. Anderson. "Presettlement Vegetation of Two Prairie Peninsula Counties." *Botanical Gazette* 140, no. 2 (1979): 232–40.
Roland de la Platière, Jean-Marie. *Encyclopedie Méthodique*. Paris: Panckoucke, 1783.
Rushforth, Brett. *Bonds of Alliance: Indigenous and Atlantic Slaveries in New France*. Chapel Hill: University of North Carolina Press, 2012.
———. "'A Little Flesh We Offer You': The Origins of Indian Slavery in New France." *William and Mary Quarterly* 60, no. 4 (2003): 777–808.
———. "Slavery, the Fox Wars, and the Limits of Alliance." *William and Mary Quarterly* 63, no. 1 (January 1, 2006): 53–80.
Russell, Edmund. "Coevolutionary History." *American Historical Review* 119, no. 5 (2014): 1514–28.
———. *Evolutionary History: Uniting History and Biology to Understand Life on Earth*. New York: Cambridge University Press, 2011.
———. "Fauna: A Prospectus for Evolutionary History." In *A Companion to American Environmental History*, edited by Douglas Cazaux Sackman, 345–74. New York: John Wiley & Sons, 2010.
———. *Greyhound Nation: A Coevolutionary History of England, 1200–1900*. New York: Cambridge University Press, 2018.
Sala, O. E., W. J. Parton, L. A. Joyce, and W. K. Lauenroth. "Primary Production of the Central Grassland Region of the United States." *Ecology* 69, no. 1 (1988): 40–45.
Sassaman, Kenneth E. "Complex Hunter-Gatherers in Evolution and History: A North American Perspective." *Journal of Archaeological Research* 12, no. 3 (September 2004): 227–80.
Sasso, Robert F. "La Crosse Region Oneota Adaptations: Changing Late Prehistoric Subsistence and Settlement Patterns in the Upper Mississippi Valley." *Wisconsin Archaeologist* 74, no. 1–4 (1993): 324–69.
Sauer, Carl O. "Grassland Climax, Fire, and Man." *Journal of Range Management* 3, no. 1 (1950): 16–21.
Sayre, Gordon M. *Les Sauvages Américains: Representations of Native Americans in French and English Colonial Literature*. Chapel Hill: University of North Carolina Press, 1997.
Schlesier, Karl H. "Rethinking the Midewiwin and the Plains Ceremonial Called the Sun Dance." *Plains Anthropologist* 35, no. 127 (1990): 1–27.

Schroeder, Sissel. "Current Research on Late Precontact Societies of the Midcontinental United States." *Journal of Archaeology Research* 12, no. 4 (2004): 311–72.

Schuler, Krysten L., David M. Leslie, James H. Shaw, and Eric J. Maichak. "Temporal-Spatial Distribution of American Bison (Bison bison) in a Tallgrass Prairie Fire Mosaic." *Journal of Mammalogy* 87, no. 3 (June 6, 2006): 539–44.

Scott, James C. *Against the Grain: A Deep History of the Earliest States.* New Haven, CT: Yale University Press, 2017.

Shackelford, Alan G. "The Illinois Indians in the Confluence Region." In *Enduring Nations,* edited by R. David Edmunds, 15–35. Urbana: University of Illinois Press, 2008.

Shay, C. Thomas. "Late Prehistoric Bison and Deer Use in the Eastern Prairie-Forest Border." *Plains Anthropologist* 23, no. 82 (1978): 194–212.

Shea, John Gilmary, ed. *Relation de la mission du Mississipi* [*sic*] *du Séminaire de Québec en 1700.* New York, 1861.

Sheehan, George. *Running and Being: The Total Experience.* New York: Simon & Schuster, 1978.

Shriver, Cam. "Thoughts on COVID Neehseehpineenki 'COVID-19' and Past Epidemics." *Aacimotaatiiyankwi* (blog), May 25, 2020. https://aacimotaatiiyankwi.org/2020/05/25/.

Silver, Peter. *Our Savage Neighbors: How Indian War Transformed Early America.* W. W. Norton, 2009.

Skinner, Alanson. "Traditions of the Iowa Indians." *Journal of American Folklore* 38, no. 150 (1925): 425–506.

Sleeper-Smith, Susan. *Indian Women and French Men: Rethinking Cultural Encounter in the Western Great Lakes.* Amherst: University of Massachusetts Press, 2001.

———. "Women, Kin, and Catholicism: New Perspectives on the Fur Trade." *Ethnohistory* 47, no. 2 (2000): 423–52.

Smalley, Andrea L. *Wild by Nature: North American Animals Confront Colonization.* Baltimore: Johns Hopkins University Press, 2017.

Smil, Vaclav. *Energies: An Illustrated Guide to the Biosphere and Civilization.* Cambridge: MIT Press, 2000.

Smith, Bruce D. *Rivers of Change: Essays on Early Agriculture in Eastern North America.* Washington, DC: Smithsonian Institution Press, 1992.

Smits, David D. "The 'Squaw Drudge': A Prime Index of Savagism." *Ethnohistory* 29, no. 4 (1982): 281.

Snyder, Christina. *Slavery in Indian Country: The Changing Face of Captivity in Early America.* Cambridge, MA: Harvard University Press, 2012.

Speth, John D. "Communal Bison Hunting in Western North America: Current Understandings and Unresolved Issues." In *"Isaac Went out to the Field": Studies in Archaeology and Ancient Cultures in Honor of Isaac Gilead,* edited by Haim Goldfus, Mayer I. Gruber, Shamir Yona, and Peter Fabian, 278–94. Oxford, UK: Archaeopress, 2019.

Stambaugh, Michael C., Daniel C. Dey, Richard P. Guyette, Hong S. He, and Joseph M. Marschall. "Spatial Patterning of Fuels and Fire Hazard across a Central U.S. Deciduous Forest Region." *Landscape Ecology* 26, no. 7 (2011): 923–35.

Stambaugh, Michael C., Richard P. Guyette, Erin R. McMurry, Edward R. Cook, David M. Meko, and Anthony R. Lupo. "Drought Duration and Frequency in the U.S. Corn Belt during the Last Millennium (AD 992–2004)." *Agricultural and Forest Meteorology* 151, no. 2 (2011): 154–62.

St. Cosme, Jean François Buisson de. "Voyage of St. Cosme, 1698–1699." In *Early Narratives of the Northwest, 1634–1699*, edited by Louise Phelps Kellogg, 337–61. New York: C. Scribner's Sons, 1917.

———. "St. Cosme to the Bishop of Quebec, 1699." In *Early Voyages Up and Down the Mississippi*, edited by John Gilmary Shea, 45–75. Albany: J. Munsell, 1861.

Stelle, Lenville J.. "History and Archaeology: The 1730 Mesquakie Fort." In *Calumet & Fleur-de-Lys: Archaeology of Indian and French Contact in the Midcontinent*, edited by John A. Walthall and Thomas E. Emerson, 265–307. Washington, DC: Smithsonian Institution Press, 1992.

Stelle, Lenville J., and Michael L. Hargrave. "Messages in a Map: French Depictions of the 1730 Meskwaki Fort." *Historical Archaeology* 47, no. 4 (2013): 23–44.

Stewart, Omer C. "Burning and Natural Vegetation in the United States." *Geographical Review* 41, no. 2 (1951): 317–20.

———. *Forgotten Fires: Native Americans and the Transient Wilderness.* Edited by Henry T. Lewis and M. Kat Anderson. Norman: University of Oklahoma Press, 2009.

Struever, Stuart. "Woodland Subsistence-Settlement Systems in the Lower Illinois Valley." In *New Perspectives in Archaeology*, edited by Lewis R. Binford and Sally R. Binford, 285–312. Chicago: Aldine, 1968.

Styles, Bonnie W., and R. Bruce McMillan. "Archaic Faunal Exploitation in the Prairie Peninsula and Surrounding Regions of the Midcontinent." In *Archaic Societies: Diversity and Complexity across the Midcontinent*, edited by Thomas E. Emerson, Dale L. McElrath, and Andrew C. Fortier, 39–80. Albany: SUNY Press, 2012.

Sutter, Paul S. "Nature's Agents or Agents of Empire?: Entomological Workers and Environmental Change during the Construction of the Panama Canal." *Isis* 98, no. 4 (2007): 724–54.

———. "The World with Us: The State of American Environmental History." *Journal of American History* 100, no. 1 (2013): 94–119.

Sutterfield, Joshua. "Aciipihkahki: Iši Kati Mihtohseeniwiyankiwi Myaamionki/ Roots of Place: Experiencing a Miami Landscape." Master's thesis, Miami University, 2009.

Swann, C., P. C. Jackman, M. J. Schweickle, and S. A. Vella. "Optimal Experiences in Exercise: A Qualitative Investigation of Flow and Clutch States." *Psychology of Sport and Exercise* 40 (2019): 87–98.

Swedlund, Alan C. "Contagion, Conflict, and Captivity in Interior New England: Native American and European Contacts in the Middle Connecticut River Valley of Massachusetts, 1640–2004." In *Beyond Germs: Native Depopulation in North America*, edited by Catherine M. Cameron, Paul Kelton, and Alan C. Swedlund, 146–73. Tucson: University of Arizona Press, 2015.

Tankersley, Kenneth B. "Bison and Subsistence Change: The Protohistoric Ohio Valley and Illinois Valley Connection." In *Long-Term Subsistence Change in Prehistoric North America*, edited by Dale Croes, Rebecca Hawkins, and Barry Isaac, 103–30. Greenwich, CT: JAI, 1992.

———. "Bison Exploitation by Late Fort Ancient Peoples in the Central Ohio River Valley." *North American Archaeologist* 7, no. 4 (1986): 289–303.

Tankersley, Kenneth B., and Nichelle Lyle. "Holocene Faunal Procurement and Species Response to Climate Change in the Ohio River Valley:" *North American Archaeologist* 40, no. 4 (2019): 192–235.

Tanner, Helen Hornbeck, and Adele Hast. *Atlas of Great Lakes Indian History*. Norman: University of Oklahoma Press, 1987.

Thomson, Betty Flanders. *The Shaping of America's Heartland: The Landscape of the Middle West*. Boston: Houghton Mifflin, 1977.

Thrush, Coll. *Native Seattle: Histories from the Crossing-Over Place*. Seattle: University of Washington Press, 2008.

Thwaites, Reuben Gold. *Early Western Travels, 1748–1846*. 32 vols. Cleveland: Arthur H. Clark, 1904.

———, ed. *The Jesuit Relations and Allied Documents: Travels and Explorations of the Jesuit Missionaries in New France, 1610–1791*. 73 vols. Cleveland: Burrows Brothers, 1896.

Tobey, Ronald C. *Saving the Prairies: The Life Cycle of the Founding School of American Plant Ecology, 1895–1955*. Berkeley: University of California Press, 1981.

Todd, Zoe. "An Indigenous Feminist's Take on the Ontological Turn: 'Ontology' Is Just Another Word for Colonialism." *Journal of Historical Sociology* 29, no. 1 (March 2016): 4–22.

Transeau, Edgar Nelson. "The Prairie Peninsula." *Ecology* 16, no. 3 (1935): 423–37.

Truteau, Jean-Baptiste. *A Fur Trader on the Upper Missouri: The Journal and Description of Jean-Baptiste Truteau, 1794–1796*. Edited by Raymond J. DeMallie, Douglas R. Parks, and Robert Vézina. Translated by Mildred Mott Wedel, Raymond J. DeMallie, and Robert Vézina. Lincoln: University of Nebraska Press, 2017.

Turnbaugh, William A. "Calumet Ceremonialism as a Nativistic Response." *American Antiquity* 44, no. 4 (1979): 685–91.

Turnbull, William D. "Late Holocene Bison Remains from Ottawa, Illinois." In *Records of Early Bison in Illinois*, edited by R. Bruce McMillan, 1–8. Springfield: Illinois State Museum, 2006.

Vale, Thomas. *Fire, Native Peoples, and the Natural Landscape*. Washington, DC: Island, 2002.

Van Nest, Julieann. "Late Quaternary Geology, Archeology and Vegetation in West-Central Illinois: A Study in Geoarcheology." PhD diss., University of Iowa, 1997.

Vergunst, Jo Lee, and Tim Ingold, eds. *Ways of Walking: Ethnography and Practice on Foot*. Burlington, VT: Routledge, 2008.

Vickers, J. Rod. "Seasonal Round Problems on the Alberta Plains." *Canadian Journal of Archaeology/Journal Canadien d'Archéologie* 15 (1991): 55.

Vinton, Mary, David C. Hartnett, Elmer Finck, and John Briggs. "Interactive Effects of Fire, Bison (Bison bison) Grazing and Plant Community Composition in Tallgrass Prairie." *American Midland Naturalist* 129 (1993): 10.

Wallace, L. L., M. G. Turner, W. H. Romme, R. V. O'Neill, and Yegang Wu. "Scale of Heterogeneity of Forage Production and Winter Foraging by Elk and Bison." *Landscape Ecology* 10, no. 2 (1995): 75–83.

Walthall, John A., F. Terry Norris, and Barbara D. Stafford. "Woman Chief's Village: An Illini Winter Hunting Camp." In *Calumet and Fleur-de-Lys: Archaeology of Indian and French Contact in the Mid-Continent*, edited by John A. Walthall and Thomas E. Emerson, 129–53. Washington, DC: Smithsonian Institution Press, 1992.

Warren, Louis S. "The Nature of Conquest: Indians, Americans, and Environmental History." In *A Companion to American Indian History*, edited by Philip J. Deloria and Neal Salisbury, 287–306. Hoboken, NJ: John Wiley & Sons, 2007.

Warren, Stephen. *The Worlds the Shawnees Made: Migration and Violence in Early America*. Chapel Hill: University of North Carolina Press, 2014.

Webb, Thompson. "The Past 11,000 Years of Vegetational Change in Eastern North America." *BioScience* 31, no. 7 (1981): 501–6.

Webb, Thompson, Bryan Shuman, and John W. Williams. "Climatically Forced Vegetation Dynamics in Eastern North America during the Late Quaternary Period." *Developments in Quaternary Sciences*, 1 (2003): 459–78.

Wedel, Waldo Rudolph. "The Central North American Grassland: Man-Made or Natural?." *Social Science Monographs* 3 (1957): 39–69.

Weisiger, Marsha. *Dreaming of Sheep in Navajo Country*. Seattle: University of Washington Press, 2011.

Welsh, Peter C. *Tanning in the United States to 1850: A Brief History*. Washington, DC: Smithsonian Institution Press, 1964.

Weyhing, Richard. "'Gascon Exagerrations': The Rise of Antoine Laumet (dit de Lamothe, Sieur de Cadillac), the Foundation of Detroit, and the Origins of the Fox Wars." In *French and Indians in the Heart of North America, 1630–1815*, edited by Robert Englebert and Guillaume Teasdale, 77–112. East Lansing: Michigan State University Press, 2013.

———. "Le Sueur in the Sioux Country: Rethinking France's Indian Alliances in the Pays d'en Haut." *Atlantic Studies* 10, no. 1 (2013): 35–50.

WHC (Wisconsin Historical Collections). *Collections of the State Historical Society of Wisconsin*. 20 vols. Madison: Madison State Printer, 1888–1931.

White, A. J., Samuel E. Muñoz, Sissel Schroeder, and Lora R. Stevens. "After Cahokia: Indigenous Repopulation and Depopulation of the Horseshoe Lake Watershed, AD 1400–1900." *American Antiquity* 85, no. 1 (2020): 1–16.

White, A. J., Lora R. Stevens, Varenka Lorenzi, Samuel E. Muñoz, Sissel Schroeder, Angelica Cao, and Taylor Bogdanovich. "Fecal Stanols Show Simultaneous Flooding and Seasonal Precipitation Change Correlate with Cahokia's Population

Decline." *Proceedings of the National Academy of Sciences of the United States of America* 116, no. 12 (2019): 5461–66.
White, John. "A Review of the American Bison in Illinois with an Emphasis on Historical Accounts." Urbana, IL: Nature Conservancy, 1996.
White, Richard. "Animals and Enterprise." In *The Oxford History of the American West*, edited by Clyde A. Milner, Carol A. O'Connor, and Martha A. Sandweiss, 237–73. New York: Oxford University Press, 1994.
———. "'Are You an Environmentalist or Do You Work for a Living?': Work and Nature." In *Uncommon Ground: Rethinking the Human Place in Nature*, edited by William Cronon. New York: W. W. Norton, 1996.
———. *The Middle Ground: Indians, Empire, and Republics in the Great Lakes Region, 1650–1815*. New York: Cambridge University Press, 1991.
———. *The Organic Machine*. New York: Hill and Wang, 1995.
———. *The Roots of Dependency: Subsistence, Environment, and Social Change among the Choctaws, Pawnees, and Navajos*. Lincoln: University of Nebraska Press, 1983.
White, Sam. *A Cold Welcome: The Little Ice Age and Europe's Encounter with North America*. Cambridge, MA: Harvard University Press, 2017.
Widga, Chris. "Bison, Bogs, and Big Bluestem: The Subsistence Ecology of Middle Holocene Hunter-Gatherers in the Eastern Great Plains." PhD diss., University of Kansas, 2007.
———. "Niche Variability in Late Holocene Bison: A Perspective from Big Bone Lick, KY." *Journal of Archaeological Science* 33, no. 9 (2006): 1237–55.
Widga, Chris, and John White. "An Ecological History of Bison in Illinois over the Last 9,000 Years." Poster presented at the Wild Things Conference, Chicago, January 31, 2015.
Williams, John W., Bryan Shuman, and Patrick J. Bartlein. "Rapid Responses of the Prairie-Forest Ecotone to Early Holocene Aridity in Mid-Continental North America." *Global and Planetary Change* 66, no. 3–4 (2009): 195–207.
Wilson, Michael, and Leslie B. Davis. "Epilogue: Retrospect and Prospect in the Man-Bison Paradigm." *Plains Anthropologist* 23, no. 82 (1978): 312–35.
Winterhalder, Bruce P. "Concepts in Historical Ecology: The View from Evolutionary Ecology." In *Historical Ecology: Cultural Knowledge and Changing Landscapes*, edited by Carole L. Crumley, 17–42. Santa Fe, NM: School of American Research Press, 1994.
Witgen, Michael. *An Infinity of Nations: How the Native New World Shaped Early North America*. Philadelphia: University of Pennsylvania Press, 2012.
Wohl, Ellen E. *Wide Rivers Crossed: The South Platte and the Illinois of the American Prairie*. Boulder: University Press of Colorado, 2013.
Worster, Donald. *Nature's Economy: A History of Ecological Ideas*. Cambridge: Cambridge University Press, 1994.
Wright, H. E. "The Dynamic Nature of Holocene Vegetation: A Problem in Paleoclimatology, Biogeography, and Stratigraphic Nomenclature1." *Quaternary Research* 6, no. 4 (December 1976): 581–96.

———. "History of the Prairie Peninsula." In *The Quaternary of Illinois; a Symposium in Observance of the Centennial of the University of Illinois*, 78–88. Urbana: University of Illinois, College of Agriculture, 1968.

———. "Patterns of Holocene Climatic Change in the Midwestern United States." *Quaternary Research* 38, no. 1 (1992): 129–34.

Wright, H. E., T. Webb III, and E. J. Cushing. "Holocene Changes in the Vegetation of the Midwest." In *Late-Quaternary Environments of the United States*, vol. 2, edited by H. E. Wright and Stephen C. Porter, 142–66. Minneapolis: University of Minnesota Press, 1983.

Zappia, Natale A. "Indigenous Borderlands: Livestock, Captivity, and Power in the Far West." *Pacific Historical Review* 81, no. 2 (May 1, 2012): 193–220.

———. *Traders and Raiders: The Indigenous World of the Colorado Basin, 1540–1859*. Chapel Hill: University of North Carolina Press, 2014.

Zedeño, Maria Nieves, Jesse A. M. Ballenger, John R. Murray, Guy Bar-Oz, Dani Nadel, Lucas Bueno, Jonathan Driver, et al. "Landscape Engineering and Organizational Complexity among Late Prehistoric Bison Hunters of the Northwestern Plains." *Current Anthropology* 55, no. 1 (February 2014): 23–58.

INDEX

Page numbers in *italics* indicate illustrations

WEYERHAEUSER ENVIRONMENTAL BOOKS

People of the Ecotone: Environment and Indigenous Power at the Center of Early America, by Robert Michael Morrissey

Charged: A History of Batteries and Lessons for a Clean Energy Future, by James Morton Turner

Wetlands in a Dry Land: More-Than-Human Histories of Australia's Murray-Darling Basin, by Emily O'Gorman

Seeds of Control: Japan's Empire of Forestry in Colonial Korea, by David Fedman

Fir and Empire: The Transformation of Forests in Early Modern China, by Ian M. Miller

Communist Pigs: An Animal History of East Germany's Rise and Fall, by Thomas Fleischman

Footprints of War: Militarized Landscapes in Vietnam, by David Biggs

Cultivating Nature: The Conservation of a Valencian Working Landscape, by Sarah R. Hamilton

Bringing Whales Ashore: Oceans and the Environment of Early Modern Japan, by Jakobina K. Arch

The Organic Profit: Rodale and the Making of Marketplace Environmentalism, by Andrew N. Case

Seismic City: An Environmental History of San Francisco's 1906 Earthquake, by Joanna L. Dyl

Smell Detectives: An Olfactory History of Nineteenth-Century Urban America, by Melanie A. Kiechle

Defending Giants: The Redwood Wars and the Transformation of American Environmental Politics, by Darren Frederick Speece

The City Is More Than Human: An Animal History of Seattle, by Frederick L. Brown

Wilderburbs: Communities on Nature's Edge, by Lincoln Bramwell

How to Read the American West: A Field Guide, by William Wyckoff

Behind the Curve: Science and the Politics of Global Warming, by Joshua P. Howe

Whales and Nations: Environmental Diplomacy on the High Seas, by Kurkpatrick Dorsey

Loving Nature, Fearing the State: Environmentalism and Antigovernment Politics before Reagan, by Brian Allen Drake

Pests in the City: Flies, Bedbugs, Cockroaches, and Rats, by Dawn Day Biehler

Tangled Roots: The Appalachian Trail and American Environmental Politics, by Sarah Mittlefehldt

Vacationland: Tourism and Environment in the Colorado High Country, by William Philpott

Car Country: An Environmental History, by Christopher W. Wells

Nature Next Door: Cities and Trees in the American Northeast, by Ellen Stroud

Pumpkin: The Curious History of an American Icon, by Cindy Ott

The Promise of Wilderness: American Environmental Politics since 1964, by James Morton Turner

The Republic of Nature: An Environmental History of the United States, by Mark Fiege

A Storied Wilderness: Rewilding the Apostle Islands, by James W. Feldman

Iceland Imagined: Nature, Culture, and Storytelling in the North Atlantic, by Karen Oslund

Quagmire: Nation-Building and Nature in the Mekong Delta, by David Biggs

Seeking Refuge: Birds and Landscapes of the Pacific Flyway, by Robert M. Wilson

Toxic Archipelago: A History of Industrial Disease in Japan, by Brett L. Walker

Dreaming of Sheep in Navajo Country, by Marsha L. Weisiger

Shaping the Shoreline: Fisheries and Tourism on the Monterey Coast, by Connie Y. Chiang

The Fishermen's Frontier: People and Salmon in Southeast Alaska, by David F. Arnold

Making Mountains: New York City and the Catskills, by David Stradling

Plowed Under: Agriculture and Environment in the Palouse, by Andrew P. Duffin

The Country in the City: The Greening of the San Francisco Bay Area, by Richard A. Walker

Native Seattle: Histories from the Crossing-Over Place, by Coll Thrush

Drawing Lines in the Forest: Creating Wilderness Areas in the Pacific Northwest, by Kevin R. Marsh

Public Power, Private Dams: The Hells Canyon High Dam Controversy, by Karl Boyd Brooks

Windshield Wilderness: Cars, Roads, and Nature in Washington's National Parks, by David Louter

On the Road Again: Montana's Changing Landscape, by William Wyckoff

Wilderness Forever: Howard Zahniser and the Path to the Wilderness Act, by Mark Harvey

The Lost Wolves of Japan, by Brett L. Walker

Landscapes of Conflict: The Oregon Story, 1940-2000, by William G. Robbins

Faith in Nature: Environmentalism as Religious Quest, by Thomas R. Dunlap

The Nature of Gold: An Environmental History of the Klondike Gold Rush, by Kathryn Morse

Where Land and Water Meet: A Western Landscape Transformed, by Nancy Langston

The Rhine: An Eco-Biography, 1815-2000, by Mark Cioc

Driven Wild: How the Fight against Automobiles Launched the Modern Wilderness Movement, by Paul S. Sutter

George Perkins Marsh: Prophet of Conservation, by David Lowenthal

Making Salmon: An Environmental History of the Northwest Fisheries Crisis, by Joseph E. Taylor III

Irrigated Eden: The Making of an Agricultural Landscape in the American West, by Mark Fiege

The Dawn of Conservation Diplomacy: U.S.-Canadian Wildlife Protection Treaties in the Progressive Era, by Kirkpatrick Dorsey

Landscapes of Promise: The Oregon Story, 1800-1940, by William G. Robbins

Forest Dreams, Forest Nightmares: The Paradox of Old Growth in the Inland West, by Nancy Langston

The Natural History of Puget Sound Country, by Arthur R. Kruckeberg

WEYERHAEUSER ENVIRONMENTAL CLASSICS

Debating Malthus: A Documentary Reader on Population, Resources, and the Environment, edited by Robert J. Mayhew

Environmental Justice in Postwar America: A Documentary Reader, edited by Christopher W. Wells

Making Climate Change History: Documents from Global Warming's Past, edited by Joshua P. Howe

Nuclear Reactions: Documenting American Encounters with Nuclear Energy, edited by James W. Feldman

The Wilderness Writings of Howard Zahniser, edited by Mark Harvey

The Environmental Moment: 1968-1972, edited by David Stradling

Reel Nature: America's Romance with Wildlife on Film, by Gregg Mitman

DDT, Silent Spring, and the Rise of Environmentalism, edited by Thomas R. Dunlap

Conservation in the Progressive Era: Classic Texts, edited by David Stradling

Man and Nature: Or, Physical Geography as Modified by Human Action, by George Perkins Marsh

A Symbol of Wilderness: Echo Park and the American Conservation Movement, by Mark W. T. Harvey

Tutira: The Story of a New Zealand Sheep Station, by Herbert Guthrie-Smith

Mountain Gloom and Mountain Glory: The Development of the Aesthetics of the Infinite, by Marjorie Hope Nicolson

The Great Columbia Plain: A Historical Geography, 1805-1910, by Donald W. Meinig

CYCLE OF FIRE

Fire: A Brief History, second edition, by Stephen J. Pyne

The Ice: A Journey to Antarctica, by Stephen J. Pyne

Burning Bush: A Fire History of Australia, by Stephen J. Pyne

Fire in America: A Cultural History of Wildland and Rural Fire, by Stephen J. Pyne

Vestal Fire: An Environmental History, Told through Fire, of Europe and Europe's Encounter with the World, by Stephen J. Pyne

World Fire: The Culture of Fire on Earth, by Stephen J. Pyne

ALSO AVAILABLE:

Awful Splendour: A Fire History of Canada, by Stephen J. Pyne

www.ingramcontent.com/pod-product-compliance
Lightning Source LLC
LaVergne TN
LVHW040908270525
812182LV00005B/308
9780295750880